D1558813

The Recovery Process In Damaged Ecosystems

Edited by

John Cairns, Jr.

University Distinguished Professor and Director
Biology Department and Center for Environmental Studies
Virginia Polytechnic Institute and State University
Blacksburg, Virginia

Second Printing, 1980

230 Collingwood, P. O. Box 1425, Ann Arbor, Michigan 48106

Library of Congress Catalog Card No. 79-89721
ISBN 0-250-40337-4

Manufactured in the United States of America

PREFACE

Change is one characteristic of an ecosystem. Species composition, various rate processes, degree of complexity and the like all vary with time. Displacements in ecosystem structure and function may result from fire, floods, glaciation, changes in rainfall and a variety of other natural causes. Additionally, displacements may result from societal stresses such as toxic chemicals, clear cutting, dams and erosion. Ecosystems may recover from both types of displacement, although the recovery process will rarely produce a system identical to the original when societal stress is involved. Case histories of recovery from human-induced displacements exist for an array of ecosystems. Since displacement results when an environmental change occurs for which the organisms are unprepared, one might assume that the recovery process would be the same for both natural and societal stresses. It is also possible that the types of stresses produced by an industrial society inhibit recolonization more than naturally occurring stresses for which evolutionary adjustment has been possible. The symposium from which this volume emanates was organized to explore some characteristics of the recovery process. Since it was limited to one day of the Ecological Society of America's portion of the American Institute of Biological Sciences annual meeting in Athens, Georgia, both breadth and depth were necessarily limited. The purpose of the symposium was to stimulate interest in restoring damaged ecosystems. The ESA meeting was an appropriate place to hold the symposium because without a sound ecological basis, restoration efforts will surely be less effective. All editor and author royalties are being given to the ESA to further the development of sound ecological concepts upon which a restoration management program can be based.

John Cairns, Jr.

Dr. John Cairns, Jr. is University Distinguished Professor in the Biology Department and Director of the University Center for Environmental Studies at Virginia Polytechnic Institute and State University, Blacksburg. He received his PhD in Zoology and his MS in Protozoology from the University of Pennsylvania, his AB in Biology from Swarthmore College, and completed a postdoctoral course in isotope methodology at Hahnemann Medical College, Philadelphia. He was Curator of Limnology at the Academy of Natural Sciences of Philadelphia for 18 years, and has taught at various universities and biological laboratories.

He received the Charles B. Dudley Award in 1978 for excellence in publications from the American Society for Testing and Materials, and the Presidential Commendation in 1971. A member of many professional societies, he is Chairman, Committee on Ecotoxicology, of the National Research Council. Dr. Cairns has been consultant and researcher for the government and private industries, and has served on numerous scientific committees. His more than 500 publications include 31 books and monographs, numerous chapters in books edited by others, scientific papers, abstracts, book reviews and congressional testimony.

ACKNOWLEDGMENTS

I wish to thank Dr. Robert Burgess, Program Chairman for the Ecological Society of America, for counsel and advice, and Dr. George Woodwell, President of the Ecological Society of America, for his introductory remarks at the symposium. The editor and authors have contributed the royalties from this book to a special discretionary fund for the Ecological Society of America to use as the Society decides.

I am indebted to Darla Donald for the many organizational and editorial duties rendered during the planning of the symposium and the publication of this volume.

CONTENTS

One of the keys to understanding the recovery process in damaged ecosystems is the successional sequences that result and their relationship to normal successional processes. Does societal perturbation merely displace an ecosystem to an earlier and more primitive successional stage or does it initiate succession of a markedly different type than that occurring naturally? Is the successional pattern or sequence following different kinds of societal perturbation as consistent and predictable as those following natural perturbations? From a management standpoint, it is important to know whether or not the successional process can be fixed in the early stages of recovery, so that the system will be more or less self-directed, or whether continual management is necessary in order to achieve the desired results. This is an extremely important determination because it involves the responsibility of companies engaged in surface mining (or that have been responsible for catastrophic spills of hazardous materials or long-term contamination of a particular ecosystem) in rehabilitating it to a more socially and ecologically acceptable condition. It would be naive to assume that all these questions can be answered in a single symposium or a single volume, especially since research on these questions has been limited. However, it is important to analyze what we know of the literature before proceeding with additional research.

Certain segments of the environmental movement tend to view environmental protection as the maintenance of ecosystems in a steady state based on present conditions. This is particularly true

INTRODUCTION

John Cairns, Jr.
Biology Department and Center for Environmental Studies
Virginia Polytechnic Institute and State University
Blacksburg, Virginia 24061

The successful restoration of the Thames River in England, Lake Washington in the United States and similar, but less publicized, efforts elsewhere have intensified interest in the recovery process in damaged ecosystems. Similarly, legislative constraints are promoting an improved definition of recovery. Virtually every recent piece of environmental legislation has specified mitigation of impacts or enhancement of available resources. Also, the prospect of large-scale surface mining to replace oil with coal as an energy base has focused attention on the need to restore strip-mined lands. The existence of machines that can remove 200 ft of "overburden" to permit the extraction of a 3-ft coal seam; the development of shears that can sever a tree trunk in seconds, enabling one to remove an entire forest quickly; the intrusion of toxic chemicals into ecosystems; and other consequences of technological development ensure that many ecosystems will be altered. Good management can reduce the severity of the impact and the recovery time. My own interest in case histories on this subject resulted in a symposium volume [Cairns et al. 1977]. The case histories in that book provide compelling evidence that both land and water ecosystems recover and that the basic recolonization process is remarkably similar in ecologically different systems. Some fundamental questions still remain:

1. How is recovery defined?
2. What criteria are important in measuring recovery?
3. Do societal perturbations (e.g., strip mining) have a different effect upon natural communities than natural perturbations (e.g., floods)?
4. Should the term "restoration" include, for example, replacement of strip-mined forests with prairie grasslands where the latter do not naturally occur (e.g., West Virginia)?
5. How do perturbation-dependent communities differ from other communities relative to the recovery process?
6. Are certain species likely to be primary colonizers of all disturbed systems?

Even the selection of an appropriate word to describe the overall process is difficult. One might use *rejuvenate*, *restore*, *renew*, *regenerate*, *rebuild* or *reconstitute*. All of these imply a return to the original (or a new) state or recreation of a youthful strength. Such terms are synonyms for "restoration" and may make it easier for readers of ecological literature to understand and identify the results expected from the recovery process.

Such discussion points up the most important questions for ecologists: "what do we wish to accomplish with polluted rivers or strip-mined lands?" For the latter this is especially important since we know which areas will be altered by future mining activities. Is it possible and economically feasible to return such systems to an earlier state following major disturbance or should we be content with an ecosystem that meets certain ecological specifications not necessarily derived from the original state? In other words, is a pond, where none existed previously, or grasslands, where there were forests, acceptable restoration? This is a particularly important question because economics may be a major factor in such decisions. For example, in the Appalachian Mountains, restoring strip-mined areas to steep, forested slopes is often much more difficult and costly than producing comparatively flat grasslands. There is evidence that the latter may furnish suitable habitat for various bird species such as horned larks *(Eremophila alpestris)*, eastern meadowlarks *(Sturnella magna)*, grasshopper sparrows *(Ammodramus savannarum)*, savannah sparrows *(Passerculus sandwichensis)* and vesper sparrows *(Pooecetes gramineus)* that historically were not common in such states as West Virginia [Whitmore and Hall 1978]. West Virginia surface mine reclamation regulations, adopted in 1971, require operators to provide adequate ground cover, minimize erosion potential and prevent acid runoff. Compliance is usually by means of hydroseeding or the spraying of an aqueous slurry of soil stabilizers (e.g., polynomial acetate), acid neutralizers, fertilizers and a seed mixture (e.g., birdsfoot trefoil and tall fescue). As a consequence, approximately 43,223 ha of new grassland have been created in West Virginia [Whitmore 1978]. This is an ecosystem unique in West Virginia but somewhat similar in avifaunal composition to the Great Plains. Whitmore [1978] also believes that effective natural succession will occur, and in 50–100 years forests will eventually reappear. Putting aside for the moment the assumption that succession will ultimately result in reforestation, a basic question is: "Should attempts be made immediately to restore ecosystems to a close approximation of their predisturbance condition?" I reviewed papers for a conference on "Surface Mining and Fish/Wildlife Needs in the Eastern United States" [Samuel et al. 1978] to write that conference summary. The primary emphasis seems to be on making a disturbed site biologically acceptable to plants and animals using vegetation that will prevent erosion. *Thus, the establishment of certain conditions (e.g., reduced erosion) have a much higher priority than the replacement of original species.*

Since it is feasible to establish more than one type of ecosystem on a disturbed site, should we use this opportunity to deliberately create a variety of ecosystems so that they will be more accessible to both the general public and those with academic interests? The University of Wisconsin has created exotic, esthetically pleasing ecosystems in Madison for educational purposes. What would be the benefits and drawbacks in making this a widespread practice when restoring damaged ecosystems? (The word "restore" in this context is used to mean "reclaim biologically.") Should ponds be placed where none existed naturally? May flatland replace mountains?

In any region, how much disturbance should be permitted at any time? A related question is: "How far should the successional process go toward the predisturbance condition before it is considered an integral part of the natural system?" This has far-reaching legal implications because it determines how long the organization which caused the damage will have management and financial responsibilities for the system. Connell's provocative paper [1978] indicates that diversity is highest when disturbances are intermediate in intensity or extent and lower at either extreme. Thus, enlightened and effective reclamation coupled with careful regulation of the size of disturbed areas and the intensity of disturbance might well increase the number of species in an area. This possibility is worth considering further in connection with surface mining and impingement or entrainment of aquatic organisms by steam electric generating plant cooling systems. Since these are ecological problems associated with energy "needs" in the United States, we will probably have to cope with them, whatever our personal views. These ecological problems are not similar to those created by chemicals at toxic concentrations, particularly when the chemicals persist and/or biomagnify.

The site of the University of Michigan Biological Station, where this introduction was written, was intensively logged near the turn of the century (Figures 1-4). The present director, David M. Gates, was here as a child and remembers the devastation, slash fires, soot and smoke. Civil engineers used the area for many years to train surveyors until revegetation made sighting more difficult, and it was given over wholly to biology. The Station is now a very pleasant environment (Figures 5 and 6), suitable for ecological research, but it is not even a close approximation of the prelogging biological condition. The area has not fully recovered if one interprets this to mean a return to original condition (Figure 7). Even if kept free from major human interference, it may never do so because of climatic and other changes. It has recovered in terms of certain uses (e.g., research, recreational or esthetic), but not others (e.g., lumbering). The stand of trees at Hartwick Pines State Park (Figure 7) provides an indication of the way things once were in parts of Michigan, and the uniqueness of that stand makes abundantly clear that the system has not recovered or been restored to its prelogging condition nor is this likely to happen in the next 50 years. Curry [1977] notes that the island of Peleiu in the Palan group of the Carolina Islands of the Pacific Ocean has not been able to reestablish the agricultural practices and patterns of pre-World War II because the damage

Figure 1. Devastation following logging near Pellston, Michigan.

Figure 2. Present site of University of Michigan Biological Station, South Fishtail Bay, Douglas Lake, near Pellston, Michigan, shortly after logging.

Figure 3. Mill at Pellston processing logs.

Figure 4. Student cabins constructed during regrowth period.

Figure 5. Lakefront area today.

Figure 6. Present University of Michigan Biological Station at South Fishtail Bay, Summer 1978.

Figure 7. Size of trees once probably characteristic of the area (Hartwick Pines).

to the soils from airfield construction, bombing, etc., was too great. Curry [1977] also notes the scale factor mentioned by Connell [1978]:

> When a tropical forest ecosystem is deforested in a monsoon or typhoon climate, there exists a short period of time when the critical available nutrients are capable of being leaked or carried out of the ecosystems. So long

> as the damage occurs to only a portion of the plant community, waterborne nutrients should be picked up by some component of adjacent communities and thus those nutrients can be saved and eventually cycled back to their sites of origin. But if intense bombing with associated fire or more "modern" use of herbicides is considered, then a single destructive event can yield a great and long-persistent reduction in site productivity. This will greatly limit the return of natural or agricultural ecosystems.

Another aspect of restoration of damaged ecosystems is the establishment of regional planning and management groups. Despite the ever-present danger of misuse of authority by such groups, they represent the best means of coping with widespread ecosystem damage as well as reducing the present rate of destruction [Cairns, 1975]. The need to limit the size of the disturbed area has been emphasized by both Curry and Connell. Enforcement of this limitation can probably be accomplished with substantial power. Epicenters for supplying new species must be within range—again a matter of regional planning. Management is also essential

1. to reclaim damaged areas;
2. to carry out biological-chemical-physical monitoring* to determine the accuracy of the predictions made on laboratory evidence;
3. to locate new industrial plants and other developmental activities so that environmental perturbations are minimized;
4. to determine the environmental carrying capacity for a particular region (e.g., population pressure, waste loading, etc.); and
5. to regulate use of the "commons."

This book will address certain specific questions:

1. What ecological factors produce perturbation-dependent ecosystems?
2. Can one quantify the degree of recovery?
3. Is recovery from major climatic change different from that following smaller-scale perturbations?
4. Is there a relationship between succession and the recovery process?
5. What generalizations can one make about recovery patterns from restored plant ecosystems?
6. What are the differences and similarities between ecosystem oscillations in structure and/or function caused by natural and societal perturbations?
7. Is rehabilitation and restoration of a major ecosystem (e.g., the Great Lakes) feasible?

An important subtheme which may emerge in this book is the relationship between structure and function in the restoration of biological integrity to a damaged system. Is community function (e.g., energy transfer) so intimately associated with community structure that one need only determine one to assess the other? Perhaps there is much functional redundancy in a community and the loss of a species does not mean the loss of a function. In this event, restoration of the redundancy would be a better criterion for

---

*I have found a surprisingly negative reaction to this word, stemming from association with surveillance of private citizens and certain organizations by federal agencies. It merely means "to watch or check on (a person or thing) as a monitor" (Webster's New World Dictionary of the American Language, College Edition, World Publ. Co., Cleveland and New York).

recovery than restoration of the function. Alternatively, function may be impaired, unaccompanied initially by a change in structure since pollutional or other stress may alter the functional capacity of species without eliminating them. Understanding these relationships is most important if we are to select the best criteria for determining recovery. Probably all three relationships exist in various ecosystems making necessary different criteria for recovery for different types of ecosystems.

Public concern about ecological systems seems to be greatest where indifference, shortsightedness, ignorance and greed are evident. Both social and ecological systems exhibit an amazing mixture of fragility and toughness. Despite daily evidence of their interactions, few people have studied the interplay of these two systems except when extreme malfunction occurs. Ecologists prefer pristine systems as free as possible from human influence. Developers view ecologists as obstacles to economic growth rather than a valuable source of information that will help reduce the number and severity of unpleasant ecological events. Stewardship of the earth's resources depends on improved communication between these two groups.

It may appear that reclamation of damaged ecosystems has nothing to do with conservation. However, if one views conservation as saving for future generations, reclamation may be as important as preservation. If complete recovery is indeed possible, a distant future generation may not be able to distinguish a reclaimed area from a preserved one. It is also worth noting that reclamation will be better understood and degree of recovery more precisely determined if untouched reference areas are preserved to serve as models.

ACKNOWLEDGMENTS

This was written at the University of Michigan Biological Station, Pellston, Michigan. I am indebted to Francesca J. Cuthbert for checking the names of the birds and to Gary R. Williams of Glen Ellyn, Illinois, for furnishing the photographs. I thank Darla Donald for editorial assistance and preparation of this volume for publication.

REFERENCES

Cairns, J., Jr. "Critical species, including man, within the biosphere," *Naturwissenschaften* 62(5):193–199 (1975).

Cairns, J., Jr., K. L. Dickson and E. E. Herricks. *Recovery and Restoration of Damaged Ecosystems* (Charlottesville, VA: University Press of Virginia, 1977.)

Connell, J. "Diversity in tropical rainforests and coral reefs," *Science* 199(4335):1302–1310 (1978).

Curry, R. "Reinhabiting the earth: life support and the future primitive," in J. Cairns, Jr., K. L. Dickson and E. E. Herricks, Eds., *Recovery and Restoration of Damaged Ecosystems* (Charlottesville, VA: University Press of Virginia, 1977), pp. 1–23.

Samuel, D. E., J. R. Stauffer, C. H. Hocutt and W. T. Mason, Eds. "Surface mining and fish/wildlife needs in the Eastern United States," Office of Biological Services, Fish and Wildlife Service, U.S. Department of the Interior, FWS/OBS-78/81 (1978), 386 pp.

Whitmore, R. C. "Managing reclaimed surface mines in West Virginia to promote nongame birds," in D. E. Samuel, J. R. Stauffer, C. H. Hocutt, and W. T. Mason, Eds., *Surface Mining and Fish/Wildlife Needs in the Eastern United States* (Washington, D.C.: Office of Biological Services, Fish and Wildlife Service, U.S. Department of the Interior, FWS/OBS-78/81, 1978), pp. 381–386.

Whitmore, R. C., and G. A. Hall. "Response of passerine species to a new resource: reclaimed surface mines in West Virginia," *Am. Birds* 32:6–9 (1978).

CHAPTER 1

THE RELATIONSHIP BETWEEN SUCCESSION AND THE RECOVERY PROCESS IN ECOSYSTEMS

Robert P. McIntosh
Department of Biology
University of Notre Dame
South Bend, Indiana 46556

INTRODUCTION

Succession is one of the oldest, most basic, yet still in some ways, most confounded of ecological concepts. Since its formalization as the premier ecological theory by H. C. Cowles and F. E. Clements in the early 1900s, thousands of descriptions of, commentaries about and interpretations of succession have been published and extended inconclusive controversy has been generated. Withal, no effective synthesis of the divergent observations from many different ecosystems, terrestrial and aquatic, has produced a body of laws and theories which ecologists, generally, have embraced. Repeated symposia on succession and the corollary problems of the succession concept, the community (or the ecosystem) and climax (equilibrium or stability in any of its several meanings) have not produced notable convergence of thought. Problems of conceptualization and terminology are still evident after three quarters of a century. The "new ecology," or better new ecologies, arrived on the post-World War II scene providing different approaches to the basic problems of ecology and succession; but the much sought after synthesis seems as elusive as ever. Recent years have seen a flurry of papers which have brought the entire concept of succession into question, and some purport to provide new insight, improved terminology and clearer direction for succession. For a consensus, however, we may be reduced to the elementary statement of succession provided in the answer to the riddle of the Sphinx—all this too shall pass away.

The apparent intractability and continued contradictions of the succession question, after many decades of study, lead to the suspicion that there is more involved than a straightforward, matter-of-fact, scientific consideration. This suggests that ecology, and the succession concept are in the midst of a revolutionary change, a change in paradigm, which is described by Kuhn [1970] as the way in which a scientific discipline progresses. Certainly, the terms revolution and paradigm, which are the keys to Kuhn's ideas, have appeared frequently in recent ecological papers [MacFadyen 1975; Goodman 1975; McIntosh 1975; Johnson 1977; Simberloff 1978]. One way of clarifying recent discussions concerning succession, which will be considered in this chapter, is to look at the history and the organizational pattern of ecology. Burgess [1977], in a history of the Ecological Society of America, showed that after 30 years of no growth, membership increased exponentially since 1950. Parallel to the increase in the number of ecologists, there has been increased heterogeneity as ecologists were required to face important scientific and empirical problems. The response to the challenges of the post-World War II era changed the traditional plant-animal or terrestrial-aquatic dichotomies in ecology. New alignments formed within ecology and new entries into its traditional disciplinary fold have introduced additional points of view to the intellectual framework of ecology [Levin 1976; McIntosh 1974, 1976].

These points of view bear examination in light of the current interest among historians and sociologists of science concerning the hypothesis of the "invisible college" as the basis of the organizational patterns which are associated with major advances and changes in concept (paradigm) in a scientific discipline [Crane 1972; Griffith and Mullins 1972]. This hypothesis is, appropriately enough, based on the logistic curve and suggests that there is a succession in development of a research field–first, rapid growth to a stable (normal) state, and then a decline. Tobey [1977] applied the invisible college hypothesis to the growth of plant ecology in the grasslands in the early 1900s and concluded that the development of grassland ecology, and its major paradigm, succession and climax, is consistent with it (Figure 1).

It is not the author's purpose to explore the invisible college hypothesis in detail, but some of the confusions and contradictions concerning succession and its associated problems may be better understood with a clearer view of the history and sociology of ecology. A common and idealized image of a scientific discipline is that it is universal and that there is free communication and mutual comprehension among its members. In fact it is generally familiar, and the invisible college hypothesis argues that any discipline, particularly one in a state of reorganization, is subdivided into loose networks of scientists with varying degrees of cohesiveness and continuity. In ecological parlance, it is heterogeneous or has pattern and is not in equilibrium. The criteria of such networks, according to Griffith and Mullins [1972] are:

1. Their members believe they are making major changes in concept or methodology. The word revolution is much in evidence.

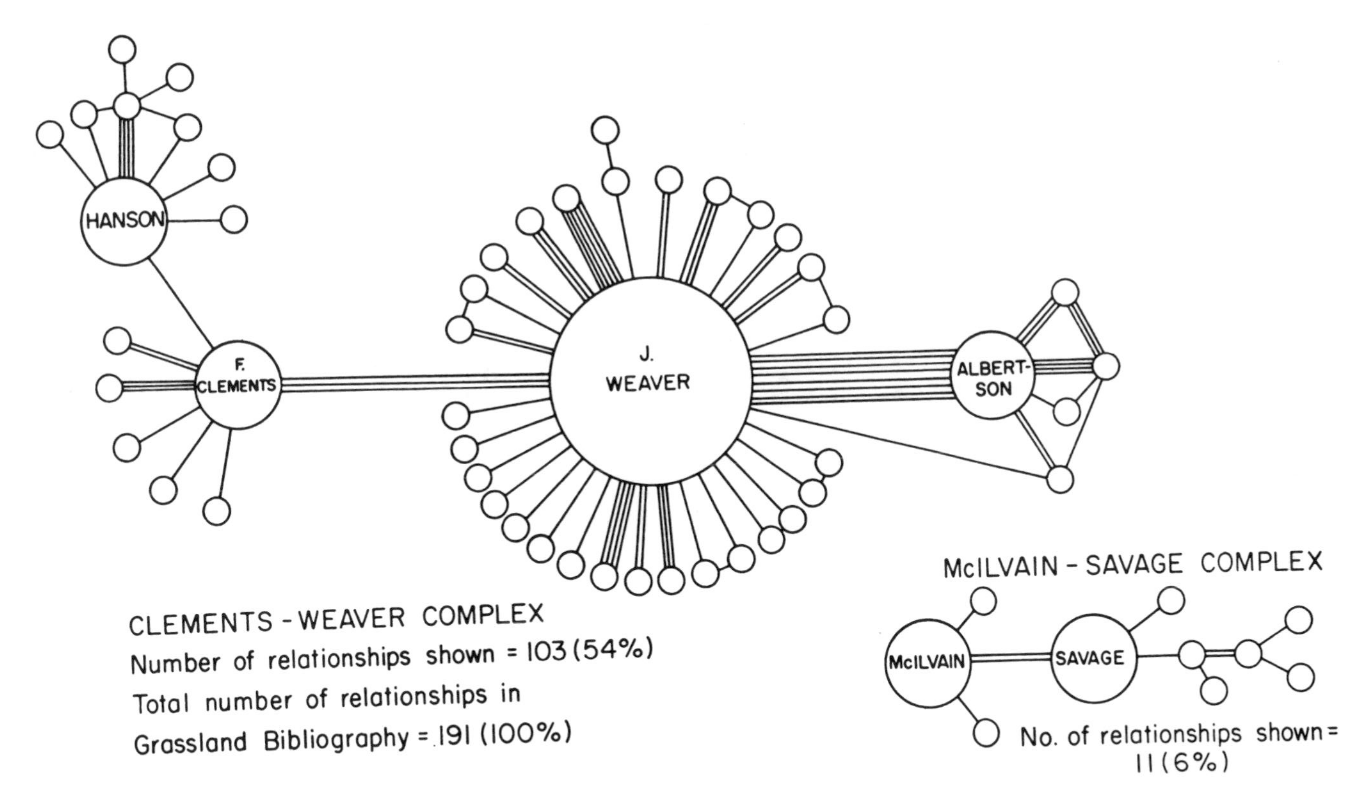

Figure 1. Major Multiple–Authorship Relationships. (Others not shown; spatial configuration for clarity only.) Source: R. Tobey, "American grassland ecology 1895–1955," in *History of American Ecology*, F. N. Egerton, Ed., Arno Press, New York (1977).

2. They do not consistently observe the attitude of disinterested objectivity desired as a norm for scientists. In fact, they may be passionate and one-sided advocates of a "ruling theory," a danger which T. C. Chamberlain [1890] warned against.
3. There is commonly a close, even somewhat closed, informal communication network within the group. This is manifest in multiple links in a citation network (cf. Figure 1).
4. Typically, a group recognizes one or more outgroups, and the more tightly organized the group is the more it sees itself as opposed to an outgroup.
5. Such groups are commonly, although not restrictively, identified with a leader, who may provide intellectual and/or organizational coherence, although these roles may be played by different individuals. There is usually a place in which the work that the group associates itself with originated, or centers on, and a more or less well defined origin and time span.

Certain developments in recent ecology suggest the existence of invisible colleges and some of the characteristics of these may be seen in the discourse on succession (Table I).

Invisible colleges are held together by identification with a common scientific problem or methodology, and in biology this may be compounded by a common taxon. In some degree they may operate as semiclosed systems and may have a somewhat parochial attitude based on a lack of familiarity with the views of other groups. Not everyone fits neatly into one or the other invisible college by virtue of academic connection or other obvious professional links; but these groups often attract "subway alumni," much as the University of Notre Dame, with which the author is affiliated, does, by virtue of the aura of success that grows about them. The author does not imply that specific assignments are definitive, and it is not necessary for the present purpose that the list be complete. The list is, of course, nationalistic in that

Table I. Tentative Invisible Colleges in Recent Ecology

Group	Leader	Place	Initial Date
Systems Ecology	Eugene Odum	U. Georgia	ca. 1950
Theoretical, Animal Community Ecology	Robert MacArthur	Princeton U.	ca. 1958
Experimental Animal Population Ecology	Thomas Park	U. Chicago	ca. 1945
Theoretical Evolutionary Ecology	Richard Levins	U. Chicago Harvard U.	ca. 1965
Plant Population Ecology	John Harper	U. C. N. Wales	ca. 1960
Quantitative Vegetation	John Curtis	U. Wisconsin	ca. 1950
Ecological Plant Physiology	Dwight Billings	Duke U.	ca. 1950

most of the persons and places cited are Americans. It omits major persons and places in other countries that may be more appropriate focal points for particular developments, e.g., R. Margalef. On a longer time scale, other individuals may clearly be seen as the initiators of some of the major facets of recent ecology, e.g., Charles Elton, G. F. Gause, G. E. Hutchinson, R. Lindeman or H. A. Gleason. The principal point the author wishes to make is the obvious one that there are several divergent positions in current ecology specifically concerned with succession, self-consciously identified with major new developments and bearing suggestive similarity with the invisible colleges of recent studies in the sociology of science. If one doubts their self-awareness, a reading of Fretwell's [1975] eulogy of Robert MacArthur or the preface of Barnard Patten's first volume on systems analysis [1971] will illustrate my point.

The concept of succession, and its corollaries community and stability, are so pervasive in ecology that any of the several invisible colleges one chooses to identify has views on them—implicit or explicit. To understand community or ecosystem stability and succession we must now consider ecological and evolutionary processes as they affect all levels of organization. A major difficulty in dealing with the problems associated with the concept is comprehending divergent approaches which have been introduced into ecology, some of which have their own, sometimes incongruent, conceptualization of the problems. Sometimes the proponents of a new point of view are unfamiliar with earlier ecological work, the thinking of other current groups, or their terminology. They may follow the lead of William Derham, a prominent physico-theologist of the early 18th Century, who purposely avoided reading the works of others so that he could himself write with more originality [Glacken 1967].

Much of the new look in succession is a form of "newspeak" [Orwell 1949], the introduction of new words for terms, ideas and phenomena long familiar to ecologists. One of the earliest words adapted for use in ecology was "pioneer," which became the standard word in the ecological lexicon to describe the early stage or stages of a sere and the life history qualities of the species occupying those stages. Clements [1905] and Gleason [1910] specified great seed production, high seed mobility, high light requirements and ability to tolerate disturbed environments as properties of pioneer species now recognized by proponents of r-selection. The term pioneer and the qualities of what Connell and Slatyer [1977] call "opportunist" species have been part of the intellectual baggage of ecologists for so long that it comes as a surprise to find pioneer placed in quotation marks by these authors, presumably as an unfamiliar term. Pioneer, as a descriptor of a successional stage and of characteristics of species, has been widely ignored in the recent rush to "fugitive," "colonizing" and "opportunistic" species—all without substantial addition to concept or fact of succession.

In keeping with this trend, many recent papers on succession replace the perfectly descriptive, long-used and apt word "disturbance" by "perturbation" (and the bastard nonword perturbate) with similarly little added insight into succession. Perturbation does not appear in the indices of the major ecological journals up to 1970. It appeared in the famous Brookhaven Symposium [1969] on stability, in Odum's 1971 textbook, in Robert May's [1973] book on stability, in Connell and Slatyer [1977] and is the preferred term in the series on systems analysis edited by Patten [1971]. It is now widely, if variously, used. In some instances it is synonymous with disturbance. In others it distinguishes natural impacts from manmade effects. Perturbations may include clear cutting of forest, fertilizing, fire or removing resident populations of rodents. The last is dignified by the name "perturbation analysis" [Schroder and Rosenzweig 1975]. Perturbation came into ecology in the late 1960s, carrying with it an aura of precision from mathematics and physics, where it has a highly specific meaning. In the physical sciences it is synonymous with small perturbation. It implies a system which differs in minor respects only from a system whose behavior is well understood and completely described—an unusual case in ecology. The substitution of such terms for long-familiar and useful terms in the ecological lexicon derives from the entry of new groups into ecology and also from the tendency of some ecologists to think that the precision, rigor and power of scientific ideas and theories from the so called "hard sciences" are readily imported into ecology along with their terminology.

The title of this chapter shares the same history. Recovery is now commonly applied to changes in community or ecosystem properties following a disturbance [Likens et al. 1978]. Its metaphorical content is powerful, particularly given the current emphasis on managed ecosystems and health of environments. In usage it is a somewhat loaded word for the traditional and well-established secondary succession which has long been defined as the sequence of change following disturbance [Golley 1977]. There is some irony in the fact that Clements, often castigated for his proliferation of ecological terms, is being upstaged by advocates of new terms. One now reads that the process of ecosystem recovery following perturbation proceeds from opportunist to mature stages.

There is, apparently, succession in the sciences—certainly there is a pecking order. Sciences are seen as developing from an immature (pioneer), fact-gathering stage toward, and ultimately to, a mature stage. Craig [1976] wrote:

> The science of ecology has matured dramatically in the last few years. From what was primarily a descriptive science has developed a new, mathematically based, evolutionary ecology.

Rosenzweig [1976] noted the immaturity of ecology as an inductive (Baconian) science and asserted:

> As sciences mature they develop a hypothetico-deductive philosophy. They progress by generating hypotheses and disproving them in controlled experiments. It is my opinion that such a maturation is now underway in ecology.

The ultimate level of scientific maturity is, of course, the capacity of prediction, and some have seen the development of ecology as a predictive science [Patten 1971; Levandowski 1972; Kolata 1972; Cody and Diamond 1975] with some cautions expressed [Cook 1974; Johnson 1977]. Harper [1977] wrote, "Plant ecology is becoming a science of vegetation management and for this it has to be a predictive science."

The model is, of course, the hard sciences, notably physics and engineering, and the means is mathematics [Patten 1971]. Various things have been introduced into ecology with the implicit assumption that they might serve as hormone injections from the more mature disciplines to stimulate the maturation of ecology. Among the earliest of these was information theory, and ecosystems [Margalef 1968]. Other recent entries into the rush to clarify changes in entropy. Succession was simply the flow of energy between ecosystems [Margaleff 1968]. Other recent entries into the rush to clarify ecology are linear and nonlinear systems theory, catastrophe theory and spectral analysis. Each of these newer entries into ecology has at least a small invisible college which is added to the established colleges in a disconcerting and unstable mixture. The track record of these in ecology is not established, although any of them can be supported by the universal justification of having heuristic merit. Information theory seems largely to have been unproductive in ecology except for providing an all-purpose measurement in the Shannon-Weaver equation [Johnson 1970]. Linear systems representations persist, although Innis [1975] said, "I do not think anyone argues any longer that biological systems are generally linear." He apparently missed Patten's [1975] statement that natural selection operates "to rid ecosystems of undesirable nonlinear characteristics. . ." The fate of the others in ecology is equally indeterminate at present. Innis [1975] also stated that "mathematics and mathematical sciences are an imposition on and threat to some of the practising biologists. In their own defence they must defame the systems approach insofar as it is identified with maths or computers." Since Innis also commented that "elderly people (past 20) have difficulty learning maths," a great many ecologists are in the defensive class. It is not clear that ecologists have been markedly unreceptive to theoretical or mathematical approaches, as Innis suggests. In fact these have become pervasive in ecology, although some proponents have claimed martyrdom in the process [Fretwell 1975; Van Valen and Pitelka 1974]. It is not beyond the pale for a practicing ecologist to require those introducing external theoretical and/or mathematical constructs into ecology to provide and recognize adequate criteria for the relevance and success of these (or lack of it) and these criteria should be ecological. One does not have to stand in opposition to "hypothetico-deductive" science to note the recurring caveats of Chamberlain [1890], Tansley [1935] and, most recently, A. J. Cain [1975]. Cain wrote: "The golden rule is always to ask questions of the animals, not of the pundits. However vociferously any particular theory in population genetics may be proclaimed, it is still necessary to ask whether it actually applies to any real populations; at present some theories appear to have some of the characteristics of religious dogma" [cf. Tansley 1935].

A BRIEF HISTORY OF SUCCESSION AND COMMUNITY CONCEPTS

A look at modern concerns about succession in the context of the history and sociology of ecology may be instructive. At the very least it may provide for new generations of ecologists an overview of the background of the continued confusion and current controversy; and it may save some reinventing of ecological wheels and redundancy in subsequent discussion and bibliographies. A number of recent frequently cited papers on succession have reviewed aspects of the subject, and some have redemolished the views on succession and climax associated primarily with Frederick E. Clements. Several provide "new" interpretations; none offers substantial new data [McCormick 1968; Drury and Nisbet 1971, 1973; Horn 1971, 1974, 1975ab, 1976; Pickett 1976, Connell and Slatyer 1977]. Golley [1977] provides a collection of reprinted articles and some useful commentary on succession; and Egerton [1977] includes reprints of several classical papers on ecology which bear rereading.

The so-called classical views of succession and the climax and supraorganismic community concepts associated with it are commonly attributed to Clements, and because of his preeminence, to plant ecologists generally. Clements' deductive and universal theories were formalized early in his career, codified in 1916, subjected to extended exegesis, and perpetuated in plant, animal and general ecological textbooks, probably because of their pedagogical convenience, long after many plant ecologists had rejected much of their substance [Egler 1951; Whittaker 1951, 1953]. They antedated, and even contradicted, many of the descriptive and early statistical analyses for which American plant ecology is commonly criticized, and are essentially a product of Clements' philosophical views [Whittaker 1953]. This is not the place to develop the details of Clements' intellectual sources; but it should be clear that his community and succession concepts were lineal descendants of a long tradition of natural philosophy which held that there were design, purpose and unity in nature and that these required a holistic approach to their understanding [Glacken 1967; Egerton 1973; Goodman 1975; Worster 1977; Simberloff in press].

Clements was neither the only nor the last proponent of this tradition in ecology; but his views fed back into philosophy, influenced Jan Christian Smuts in his formalization of the philosophy of holism, and engendered in some ecologists what was described as religious fervor [Tansley 1935]. Clements [1916] described succession as a universal, orderly process of progressive change. He asserted that the community (association *sensu* Clements) developed from diverse pioneer stages to converge on a single, stable, mesophytic community (monoclimax) under the control of the regional climate. He held that the association repeated in its development the sequence of stages of development of an individual organism from birth through death and was orderly and predictable in its development in the same sense as the development of an individual organism. As succession proceeded, the association came increasingly to control its own environment and, barring disturbance, became self perpetuating or climax. Clements regarded stabilization as essentially a synonym of succession but recognized its limits, writing, "the most stable association is never in complete equilibrium, nor is it free from disturbed areas in which secondary succession is evident." He was the coiner of the phrase "dynamic ecology," and an early proponent of the importance of function and process in succession. Until the arrival of niche theory (ca. 1957), no ground in ecology was more exhaustingly worked over than Clements' association and climax concepts. To his eternal credit he asserted in the first treatise on ecological research published in America [Clements 1905; cf. Goodman 1975] the recently rediscovered maxim that stability in succession is not strictly associated with species diversity, Clements wrote,

> The number of species is small in the initial stages, it attains a maximum in intermediate stages; and again decreases in the ultimate formation, on account of the dominance of a few species.

This, and other insights, are often overlooked in criticisms of Clements' succession theories.

Community concept and succession in America were largely the province of early plant ecologists with the notable exception of Victor Shelford, who asserted the primacy of community in animal ecology and studied successions of fish ponds and of tiger beetles in sand dune succession [Shelford 1911, 1913]. Shelford collaborated with Clements in formulating the biome concept [Clements and Shelford 1939], which, for all its inadequacies, remains the study unit of choice for modern ecosystem ecologists. The importance of synecology was also recognized by the premier British animal ecologist, C. S. Elton [1927], who wrote, "it is clear that the study of the autecology of the numbers of any species involves inevitably a consideration of the synecology of the community in which it lives." Although the concept, or "heresy" [Colinvaux 1973], of the supraorganismic community is commonly attributed largely to Clements, it was in fact, standard doctrine among animal as well as plant ecologists, most of whom shared similar natural history traditions. S. A. Forbes was a primary expositor of the concept of organization in nature, and his article, "The Lake a Microcosm" [1887], is widely cited. His earlier article [1880], "On Some Interactions of Organisms," is a less familiar but explicit statement of the organismic concept, and a prescient comment on population interactions as molding a community and the energetic basis of the ecosystem. Forbes wrote in 1883:

> A group or association of animals and plants is like a single organism in the fact that it brings to bear upon the outer world only the surplus of forces remaining after all conflicts interior to itself have been adjusted. Whatever expenditure of energy is necessary to maintain the existing internal balance amounts to so much power locked up and rendered unavailable for external use. In many groups this latent energy is so considerable and is a liability to

> such fluctuation that a knowledge of its amounts and kinds, and of the laws governing its distribution, is extremely important to one interested in measuring or forseeing the sum and character of the outward tending activities of the class.

The prominent American animal ecologists, W. C. Allee, A. E. Emerson, O. Park, T. Park and K. Schmidt [1949], authors of the major mid-20th century work on animal ecology, adopted a firmly organismic view of the community and succession. Allee's own views on cooperation among organisms were widely published and attributed by Ghiselin [1974] to his Quaker background. However, since Ghiselin puts Clements in the Chicago school of ecology and cites "Vernon" Shelford, this may not be precisely correct. In any event not all the animal ecologists with an organismic concept of community are Quakers. Alfred Emerson, who was of the Chicago school, was certainly an explicit proponent of the organismic concept and the evolving community unit which permeated the thinking of most animal ecologists [Emerson 1960]. Bodenheimer [1957] reviewed the attitudes of ecologists on community and stated that the supraorganismic concept was firmly established in animal ecology and limnology, but that its empirical bases were not well established.

Some recent commentators on succession and community attribute a rather monolithic Clementsian view to plant ecologists and point out its inadequacies as a prelude to suggesting new looks for ecologists. It is not always clear in these presentations that the Clementsian position was, from its inception, widely criticized and by no means universally accepted among plant ecologists, although it permeated the textbooks [Egler 1951]. The Clementsian, supraorganismic, holistic, community and succession concepts were probably more widely and less critically accepted by animal ecologists than plant ecologists. Animal ecologists had turned their attention in good part to physiological or population ecology, left the development of community and succession concepts to plant ecologists and readily absorbed the Clementsian positions which fitted their own preconceptions. To many plant ecologists the community was less well defined and unitary, and succession less orderly, than Clements suggested. H. A. Gleason and W. S. Cooper were the most notable American dissenters. Gleason [1917] opposed the rigid Clementsian views of orderly, progressive succession [McIntosh 1975]. He asserted, ". . . succession is an extraordinarily mobile phenomenon whose processes are not to be stated as fixed laws, but only as general principles of exceedingly broad nature and whose results need not and frequently do not, ensue in any definitely predictable way." Cooper [1913, 1923, 1926] produced numerous studies of succession and generally applied the concept of multiple seres to a climax but was not a doctrinaire follower of Clementsian concepts. Cooper [1913] wrote the climax forest is "not homogeneous throughout in character and appearance. It is made up of small patches of diverse aspect which represent stages in an endless chain of permutations, the total result of which is that the forest as a whole remains the same, although a given area is constantly changing." In effect, he described the mosaic view of climax commonly attributed to more recent ecologists. Sir Arthur Tansley [1920, 1935], the British counterpart of Clements in preeminence, even-handedly rejected Clements' extreme organismic view and Gleason's individualistic concept. He allowed, however, that the community might be compared to a "quasi-organism."

Gleasons' major contribution to ecology was his individualistic concept. This idea, advanced in opposition to the organismic concept of community and succession developed by Clements, although long ignored is now one of the most influential in ecology [McIntosh 1975]. The dichotomy initiated by Gleason is clearly antecedent, if not antebellum, to the anticipated state of ecology at the present time as later comments will show. Bodenheimer [1957] was an unusual dissident from the supraorganismic view among animal ecologists, adopting substantially Gleason's position. According to Bodenheimer:

> Within the biocommunities interactions occur between various species of plants and animals, but even the beginnings of any real integration of the members of the biotic community, the biocoenosis, into a supraorganismic structure have never been demonstrated. As a rule, each species exists within a community in its own right, which is expressed by the different territory of every species as conditioned by its own reaction basis, different from that of all other species. Little active co-operation occurs in the animal community; more often we find a certain mutual tolerance of such species, whose niches overlap partially.

Worster [1977], in a recent book on the origins of ecology, incorrectly sees the organismic tradition extending through W. C. Allee at Chicago but disappearing with his retirement. Pickett [1976] commented that the Clementsian model of succession has been abandoned by modern ecologists but no contemporary model has replaced it. In fact, the succession concept advanced by Margalef [1968] and Odum [1969] continued the substance of the traditional organismic and Clementsian concepts as various ecologists have noted. Odum noted a parallel between succession and the developmental biology of organisms which is redolent of Clements' supraorganism or of Tansley's quasi-organism. He described the ecosystem as a unit of organization undergoing an orderly process of development that is reasonably directional, and therefore predictable, community controlled and culminating in a stabilized (mature) ecosystem. The functional, or ecosystem, approach to succession provided a series of functional criteria (trends to be expected) in lieu of compositional criteria of seral stages. Clements became subject to criticism when he spelled out his successional and organismal community concepts and gave the climax association geographical extent, life form and compositional criteria. Ecosystems ecologists generally avoid this although they continue some of the traditional recognition of climax communities. Patten [1975] followed Emerson [1960] in seeing the ecosystem as an evolutionary entity. Thus, the core of the organismic position espoused by Clements and evident in most traditional animal ecology, although less explicit in specific regional or compositional detail, is still evident in ecosystem ecology.

Ghiselin [1974] said that holism has gradually merged into systems analysis. Cook [1977] notes that Raymond Lindeman's trophic-dynamic aspects of ecology emphasized the relationship of food cycle relationships to succession. According to Cook, Lindeman followed Thienemann, an early proponent of the organismic concept applied to lakes [Bodenheimer 1957]. Cook writes that Lindeman brought energy relations into the analysis of successional development: "It is here that the analogy between the development of the organism to maturity and community changes during the succession finds its fullest expression; and it is the elaboration of this metaphor which has provided continuing inspiration to community ecologists." To exemplify this Cook cites Odum [1977]. Simberloff [in press] states that the organismic concept is "not dead, but rather transmogrified into the belief that holistic study of ecosystems is the proper course for ecology." Simberloff suggests that this is an expression of philosophical tradition continuous with Greek metaphysics and certainly it is continuous with traditional ecology.

Much of the currently evident schism in ecology and the divergent views on succession are continuations and elaborations of the dichotomy which first became dramatically apparent in the contrasting views of Clements and Gleason. The divergence between individualistic and holistic approaches in ecology is manifest in much of the current literature. Connell and Slatyer [1977] comment that Odum's "trends to be expected" in succession derive from the organismic analogy. They write, "This view is based solely on the analogy, not, in our opinion, on evidence." Harper [1977] states,

> "There is an important sense in which our knowledge of terrestrial ecology has been determined by these contrasting philosophies: an individualistic interpretation based on the history of a community and a holistic interpretation that has seen a community dominated by the constraining forces of limited resource or driven towards some stable constrained state."

Eugene Odum [1977] sees what he calls the "new ecology" as dedicated to holism and dealing with the supraindividual levels of organization. Like Clements and others of his predecessors, Odum sees new systems properties emerging in the course of ecological development and a holistic strategy for "ecosystem development," which is his preferred term for succession. Succession theory had, from its earliest Clementsian origins, incorporated the concomitant and reciprocal changes (reactions) on the environment associated with vegetational change. Accumulation of organic matter, increased moisture and nutrient supply, modification of light and general physical environment were part of the traditional concept of succession. In this sense, succession has always been an ecosystem theory and leads directly to the ecosystems approach [Tansley 1935].

Odum [1977], commenting on ecology as a new, integrative discipline, notes the contrary views of reductionists in ecology and succession but does not emphasize that this invisible college also regards itself as the "new ecology." Clements and Odum converge in their descriptions of succession as an orderly, predictable, unidirectional process of change which results in modification of and control over the physical environment and culminates in a stable (mature) ecosystem. However, it is just the lack of order, direction, uniformity and predictability and the substantial differences in approach to succession that have brought us to yet another symposium. Nothing could be more neatly organized than Clements' deductive, deterministic, organismic system, and Odum's holistic, integrative "new ecology" is its lineal descendant. From Cooper and Gleason on many ecologists have found these views difficult to accept. The problems lie in the observation by Whittaker [1953] which seconds Gleason's observation quoted above:

> "Succession may thus be thought to occur, not as a series of distinct steps, but as a highly variable and irregular change of populations through time, lacking orderliness or uniformity in detail though marked by certain fairly uniform overall tendencies."

A major dichotomy or polarity is still evident in current ecology between advocates of holistic approaches to community or ecosystems and succession and proponents of reductionist, individualistic or population centered approaches [Golley 1977]. Whittaker [1975] noted the lack of a bridge between them. The invisible colleges have shifted, and new foci, personalities and lineages are all too evident. Convergence of precepts may be seen in some cases, but an embarrassing heterogeneity creates great difficulties in assessments of community and succession. What I have described as a polarity, its historical origins and present states, might be better subjected to an ordination, cluster analysis or cladistic analysis if not a Markov process.

THE "NEW" SUCCESSION OR "RECOVERY FROM PERTURBATION"

Traditional succession studies posed the question, "What is the order of species populations in time?" A 1973 conference on succession addressed the question, "Are there ordered patterns of populations in time?" Some ecosystem ecologists believe that species patterns are largely irrelevant to ecosystem changes over time. The search for regularity and order, which is the traditional hallmark of science, is not as readily pursued in ecology as some would hope. Clearly, progress has occurred since 1903 when H. C. Cowles [1904] described ecology as chaos; however, pattern recognition remains elusive. Robert Whittaker's metaphor, a shimmer of populations, is aptly descriptive, at least for those who believe that populations matter.

Recent critics of succession concepts have commonly concentrated on extreme interpretations of Clementsian succession. Some have emphasized Egler's [1954] distinction between "relay floristics" and "initial floristics" and agreed with his assessment of the latter as a "startlingly new form" in interpretation of old-field succession (Figure 2). Relay floristics (the upper diagram) simply suggests the classical assumption of amelioration of the site by sequential groups of species as unit communities, each presumably making the site unsuitable for themselves but more suitable for invasion by the next group of species. This interpretation was based largely on early studies of

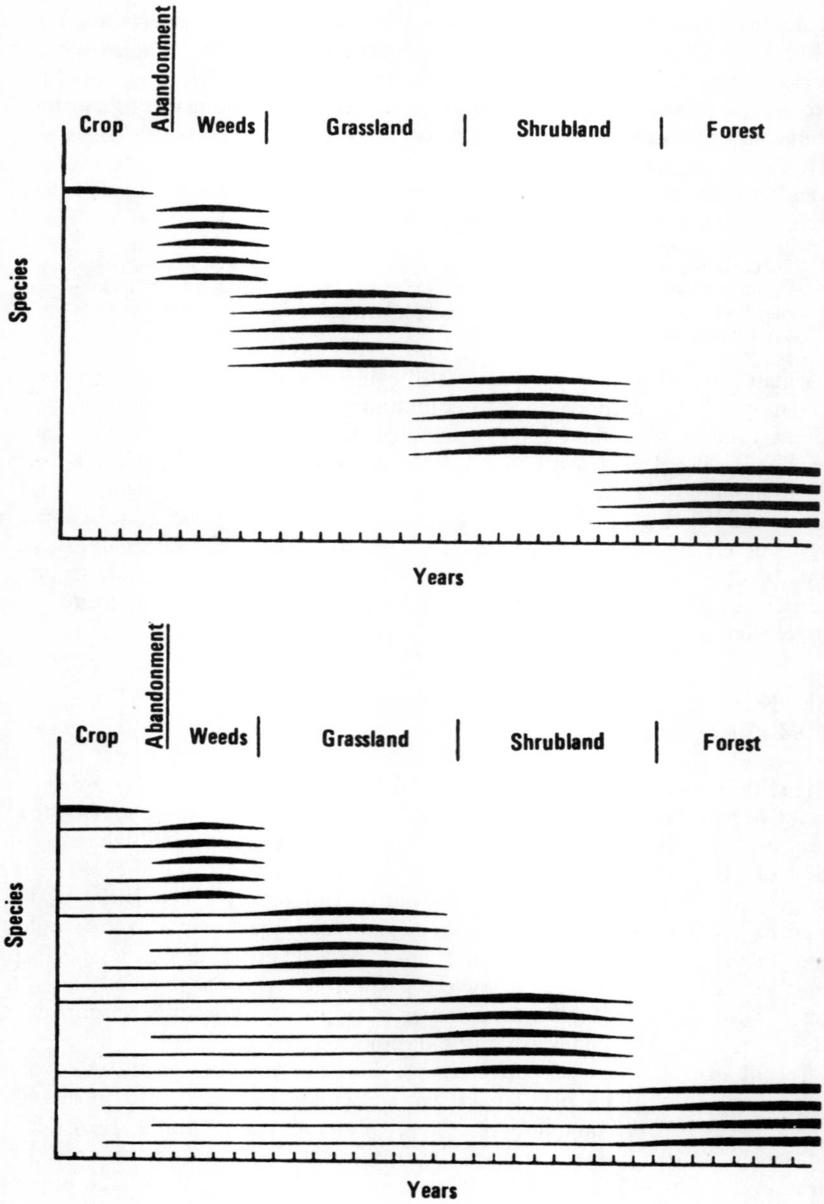

Figure 2. Upper-"relay floristics," lower-"initial floristics." Source: F. E. Egler, "Vegetation science concepts. I. Initial floristics composition, factor in old-field development," *Vegetatio* 4:412–417, Dr. W. Junk bv. Publishers, The Hague (1952–54).

primary succession [e.g., Cowles 1899, 1901] and their incorporation into Clements' organismic successional concept with its emphasis on "reaction" as the effect of organisms in modifying the environment. Relay floristics is a series of unit communities (groups of species) entering and leaving the sere, and rising to peak populations, essentially simultaneously as the members of the successive communities successfully invade the site presumably modified by their predecessors. Vegetation is to Egler, as it was to Clements, "complexes of plant communities existing in nature as organized wholes." Actually, as Egler commented, the general change is one, "of gradual alteration, in which the appearance and disappearance of any one community may be difficult, and needless, to pinpoint." Egler's assertion that the term succession necessarily connotes discrete community jumps is not true. These are assumed only in a restricted theory of succession and of the community unit. Egler's view of a gradual alteration is entirely compatible with other views of succession and is inherent in Gleason's [1917, 1926] individualistic concept and similar concepts current today which are not suggested by Egler's diagrams [McIntosh 1967; Whittaker 1951, 1953, 1967]. Egler, however, regarded succession as a "phenomenon on a high level of sociologic integration, not a matter of individual species." Hence, he preferred the term "development," as does Eugene Odum for similar reasons. Sometimes an interpretation has been placed (not necessarily by Egler) on an individualistic viewpoint which suggests that the plant is independent of effects of other plants or organisms generally, and subject only to the physical environment. That was certainly not Gleason's [1910] view nor that of his successors. Gleason wrote,

> The plant itself is in many cases the controlling agent in the environment. . . . The establishment of a plant in the place which it occupies is conditioned quite as much by the influence of other plants as by that of the physical environment.

His individualistic concept of succession was explicit. "Succession, therefore, as an ecological process, is no more than the mass effect of the action or behavior of individual plants and relates itself perfectly to the individualistic concept of vegetation . . ." [Gleason 1927].

Egler's initial floristic composition was restricted by definition to secondary succession, largely to old fields, and assumed the site to be already populated at abandonment by seeds or vegetative material of the full range of seral stages. These develop without, or with only restricted, invasion of additional species (relay floristics). Species of the initial floristic set assume dominance in sequence, in Egler's diagram, as groups of species rising and falling synchronously. If dissemules of any stage, including the last, are absent at abandonment, that stage cannot readily invade the site and will be absent, or much delayed, waiting on invasion (relay floristics). Egler emphasized the importance of initial floristics in old-field succession, but there is no reason why it should not be generalized to secondary succession at large. The principal point he made is that initial floristics explains certain relatively stable plant communities which change very slowly due to restriction on subsequent invasion of other organisms by the initial populations (Figure 2).

Views and observations on succession have varied greatly among different individuals and invisible colleges, past and present. McCormick [1968], in his admirable general review of succession, noted the views of a number of early plant ecologists on the sometimes surprising persistence of pioneer species in later seral stages and the presence of propagules on disturbed sites. It was no surprise, even to Clements [1916], that a newly bared, secondary succession site might contain viable reproductive units of diverse stages, even the climax, of the sere. He defined primary bare areas as those lacking viable germules of other than pioneer species, having no effect of prior occupation by organisms, and requiring long-term effects of organisms before they are ready for climax communities. Secondary bare areas, he said, "possess viable germules of more than one stage, often in large number, retain more or less of the preceding reactions, and consequently give rise to relatively short and simple seres." Oosting and Humphries [1940] among others recorded the persistence of viable seed in successional stages of various ages [cf. Livingston and Allessio 1968]. The possibility of residual propagules, abbreviated seres or skipped stages was clearly recognized by plant ecologists even of the old school. That secondary succession following disturbance by cutting of a forest may produce some surprises would not surprise a co-contributor to this book, Dr. Vogel. In his studies of Crex Meadows in Wisconsin [Vogel 1964] he found that when a jack pine forest was clear cut it was immediately replaced by full-blown prairie, although prairie species had not been prominent in the precut forest. They had persisted as suppressed vegetative forms up to 60 years in the forest. Converse observations of forest species developing on grassland, from grubs following cessation of fire, are very much a part of successional observations in the midwest by Gleason and many others [Curtis 1959; Auclair and Cottam 1971].

Essentially, Egler argued that an established vegetation cover resists invasion (relay floristics) from the outside—a thesis supported by seeding experiments, which have shown that even artificial seeding does not readily replace established vegetation [Sagar and Harper 1960; Juhasz-Nagy 1964, Karpov 1964]. It is not always clear in initial floristics whether the later species to appear are lying dormant through the dominant period of the earlier species or whether they are growing slowly but suppressed and presumably missed in studies of early stages. Egler stressed instances in which a later component of a sere is absent by chance from the initial floristic composition which will then allow persistence of a stable community of an earlier stage. This is a variant of the position of Gleason that the species which will occur on an available site are substantially a matter of probability depending on the availability of seeds and favorable environment. Cooper [1923] provided a diagram (Figure 3) very similar to Egler's initial floristics and noted that even in a primary succession on glacial till, the climax species, spruce and hemlock, may appear in the first year along with the pioneer species [cf. Drury and Nisbet 1973]. Cooper's diagram, unlike Egler's, shows an individualistic sequence of species rising to dominance and then diminishing. Cooper [1926], however, contrary to Gleason, believed that the community

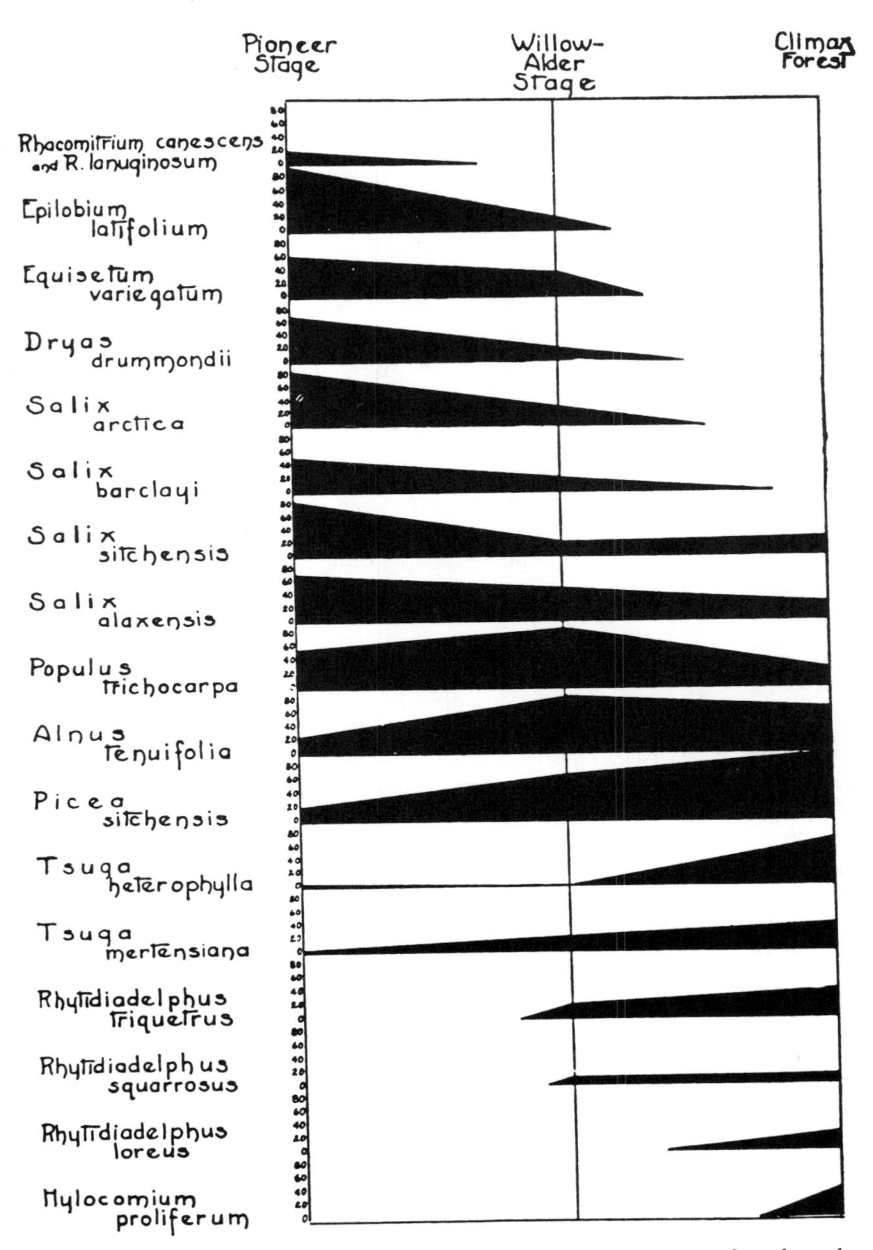

Figure 3. Successional roles of eighteen important species; percentage of total number of stations in which each was found is shown on the left of the diagram, and indicated by the thickness of the black areas. Source: W. S. Cooper, "The recent ecological history of Glacier Bay, Alaska. II. The present vegetation cycle," *Ecology* 4:223–246 (1923).

was governed by emergent properties, "so that their mass action is not equivalent to the sum of the actions of the component individuals." The fact is that early successional concepts were not all cast in the mold of the Clementsian monolith frequently attacked; there was independent assortment of ideas and conflicting interpretations among plant ecologists. Tansley [1935] raised the question of the meaning of the summation of the actions of the individuals about which Cooper and Gleason differed.

McCormick [1968] noted that classical succession theory is sometimes associated with a theory of geological base leveling, which is vigorously attacked by Drury and Nisbet [1971, 1973]. Although the meaning of Drury and Nisbet's [1971] paper, like Margalef's [1958] paper, is somewhat obscured by the necessity of fitting into the rubric of general systems, they were critical of any successional system which develops in a "preordained course to a stable equilibrium"; they attacked evenhandedly both plant and animal ecologists who urged this. Clementsian plant ecology was criticized for what they described as its analogy with an outmoded theory of geological base leveling and peneplanation—an analogy between peneplain and climax. Drury and Nisbet argued that the assumption that the landscape develops in a preordained course by geological processes to a stable equilibrium predetermines that plant and animal communities will do so. They cited "the classical concepts of cycles of erosion . . . on which the classical theory of plant ecology was based" [Drury and Nisbet 1973]. They advocated replacement of older concepts of geological erosion cycles with a noncyclic kinetic or equilibrium model of geological change. The dispute concerning the geological phenomenon of crustal stability is, according to Wright [1974], "at an impasse" not, as Drury and Nisbet asserted, resolved in favor of the new kinetic model. Whatever the merits of that dispute, it is largely irrelevant to considerations of succession which ecologists generally agree operate on a different time scale from geological base leveling. Clements' theories did not derive specifically from a deterministic view of geomorphology. In all likelihood both concepts derived, as Drury and Nisbet suggested, from a more ancient and powerful tradition of unity, stability and organization in nature extending back to the roots of Greek philosophy [Glacken 1967; Simberloff in press]. In any event, modern plant and animal ecologists do not rest their concepts of succession on geological base leveling or stable climates. Even Clements and Cowles, whose observations on sand dunes influenced Clements' successional concepts, recognized that vegetation cycles were much more rapid than the presumed geological cycles. Cowles early training in geology with Chamberlain no doubt influenced his concept of physiographic succession, but he was explicit in saying that geological factors were not primary [Cowles 1911], contrary to Wright's [1974] interpretations and Drury and Nisbet's assumptions. Cowles wrote:

> "It has been seen that changes of climate or of topography generally institute vegetative changes, indeed this would have been predicted to be the case even without examination. But at first thought it seems somewhat striking that far reaching vegetative changes take place without any obvious climatic change

and without any marked activity on the part of the ordinary erosive factors. . . . So rapid is the action of the biotic factors that not only the climate, but even the topography may be regarded as static over large *areas* for a considerable length of time."

Only the most extreme and long since outmoded Clementsian interpretation justified the development of a regional monoclimax on the assumption of geological base leveling [Tansley 1935, Whittaker 1951]. Only Braun among plant ecologists linked vegetation concepts closely to the geological cycle and recent palynological evidence contradicts her interpretation [Wright 1974].

Drury and Nisbet [1971] asserted that the classical Clementsian model excludes reciprocal effects of the plants on the environment and that, "in a rigid extension of the theory, the animals themselves will form a climax community and no interactions between animals and plants need to be considered at all." Whatever the logic of that, plant ecologists were certainly aware of the effects of the plants and animals on the environment. Clements' succession concept was in considerable part based on *reaction* [cf. Drury and Nisbet 1973] which was the effect of plants on the environment. Cowles [1911] commented on the importance to the plant geographer of "the vegetative changes that are due to plant and animal agencies." Clements, Tansley, many of their contemporaries and later plant ecologists recognized that animals could divert the course of succession, control stable communities or even cause retrogression (see below). Has any animal ecologist argued that the existence of stable, efficient animal communities is established, "by assuming that they must have done so in an ancient climax, on an ancient peneplain," as Drury and Nisbet state? Thus, it is not clear why it is necessary to dispute the point.

Drury and Nisbet's [1973] followup article is widely cited and is self described as offering an alternative explanation of succession. They posed their views as alternative to the views of Margalef, MacArthur and Connell (whom they state revived Clements' organismic concept), Odum and Whittaker. They described these authors, collectively, as attributing succession to properties of the community. This article incorporated a basic problem about what succession is and how it is demonstrated. The authors distinguished succession "in space" (zonation) and succession "in time." They commented further, ". . . to provide a unified description on which a general theory can be based, it is necessary to assume a homology between a spatial sequence of zones of vegetation visible at one time on the landscape and a long-term sequence of vegetation types on a single site." This assumption, they asserted is "reinforced by the classical geomorphological theory of landscape development" they claimed in their 1971 paper had led ecologists to view succession as a preordained sequence. For unaccountable reasons, they cite Gleason [1927] by page number as a reference for this assumption. Gleason, like many ecologists, was flatly opposed to such an assumption. He says of the time and space sequences, "never the twain shall meet," and "Areal zonation of vegetation does not constitute a sere, and is correlated with succession only in exceptional cases." It is not clear who supports the assumption that succession in any current ecological sense refers to other than a sequence in time.

Drury and Nisbet reviewed the applicability of and the evidence for "the Odum-Whittaker criteria" meaning the "trends to be expected" advanced by Odum [1969]. Whittaker's views of succession do not correspond with Odum's sufficiently to hyphenate them; Odum-Margalef would be more acceptable. It is not possible here to examine their review in detail; although it should be noted that they concentrate on secondary, primarily old-field, succession in a temperate forest region. Their summary judgment was that the evidence does not support Odum's "trends to be expected."

Drury and Nisbet offer a "sketch of an alternative explanation" for succession, "based on the observed congruences along environmental gradients" mentioned above. This "observed congruence," is transformed from their assumed homology and has been attacked as a weak link in the successional literature. It is true, as they observe, that only the earliest stages of succession in limited cases, such as old fields, have been directly observed. Many studies, particularly of primary succession, have made the assumption that sites of different age, forming zones in space, may be linked together as a putative chronological sequence [e.g., Cowles 1901; Olson 1958]. Very few have demonstrated a homology between zonation and time. The studies of McCormick and his students of primary succession on granite outcrops, discussed below, are unusual in this respect.

There are numerous difficulties in demonstrating the homology or congruence, as McCormick's studies illustrate. Some critics of succession studies have argued that only a direct observation of change in time, preferably coupled with experiments, can afford adequate evidence [Swan and Gill 1970; Austin 1977]. Hence, the concentration on old-field or microcosm studies. It is familiar that many successions do not occur in time scales convenient for human examination, and it may be nearly impossible to achieve the ideal. An obvious problem in meeting this ideal is the difficulty in finding replicates of long-term time sequences which may be used to demonstrate the presumed common properties of a sere. It is not clear how the "observed congruence" is an "explanation for most of the phenomena of succession . . . as consequences of differential growth, differential survival, (and perhaps also differential colonizing ability) of species adapted to growth at different points on environmental gradients," as stated by Drury and Nisbet.

It is difficult to find in Drury and Nisbet's interpretation of succession something new or "alternative." Their "key statements" are:

1. Rapid dispersal mechanisms and ability to tolerate physical stress of harsh environments are commonly associated.
2. Colonizing ability and growth rate tend to be correlated with size.

These are dubiously listed as "hypotheses" when they are basically empirical observations which, like most ecological observations, are more or less true. It is not clear that rapid dispersal is at a premium if initial floristics is an important aspect of succession as they intimate. Whatever the validity or novelty of Drury and Nisbet's alternative explanation, they come down firmly opposed to a supraorganismic concept of succession. "A comprehensive theory of succession should be sought at the organismic or cellular level, and not in emergent properties of communities." This may be true, but no clear-cut evidence is advanced by these authors to support the generalization nor is there effective guidance as to how ideas of succession should be integrated with current work in population theory much less with cellular level theory.

Another frequent, if somewhat redundant, contributor to the recent literature and symposia on succession is Horn [1971, 1974, 1975ab, 1976]. Horn [1974] adopted the approach "championed by Drury and Nisbet" [1973] in what he described as "first steps toward theories of succession that are based directly on properties of organisms rather than emergent properties of ecosystems," an approach that Horn characterized as "vitalistic." Horn concentrated on "recent developments in population biology that have profound implications for theories and patterns of secondary succession." He attributed secondary succession to competition; early species "producing an environment in which later species are competitively superior." This places him forthrightly in the camp of the many traditional and some recent ecologists who believe that competition is the major force organizing communities and controlling succession [Diamond 1975; cf. Weins 1977; Pulliam et al. 1977]. It is, therefore, not surprising that in two of his papers Horn commented on Diamond's classification of "supertramps" (pioneer for the birds) which Horn described as, "certainly adaptable to the study of plants in succession." Diamond [1975] developed a widely cited approach which described "assembly rules" for species combinations. The phrase is reminiscent of the early use of "valence" [Raunkaier 1939] which suggested similarly neat combinations for the occurrence of plant species. It would be pleasant to find plant species following a set of combinatorial rules, like elements, on the basis of interspecies competition. However, Connor and Simberloff [in press] have argued that Diamond's assembly rules are either tautological consequences of their definitions or describe patterns which would have resulted from random distribution on islands. What is clear is that concepts of community structure and succession, which are predicted on an assumption of competition as the organizing force, need cautious review; plants may or may not be amenable to Diamond's assembly rules or to Horn's "transition matrix."

Horn [1974] suggested the inadequacies of traditional ecologists stating that much of the conventional wisdom of succession is based on "biased definitions." He provided a definition of what he meant by succession, which turned out to be what ecologists have always meant by secondary succession. If one winnows Horn's several papers on succession, the burden of his new insight into succession is based on perhaps the most maligned aspect of Clementsian successional concept. Horn [1975a] seized on the "dramatic property of succession . . . its repeatable convergence on the same climax community from any of many different starting points" and noted its analogy to a statistical process—the Markov chain. From this observation he moved quickly to the recognition that "Several properties of succession are direct statistical consequences of plant-by-plant replacement process and have no uniquely biological basis." The vitalistic interpretation of succession Horn had seen was replaced by a statistical one leaving "no unique biological basis" Horn [1975a].

Horn saw a "convergence" of a number of recent workers on the Markov model which, in his case, rests upon the assumption that the plant-by-plant replacement probabilities can be determined from the presence of saplings under adult trees. Horn's [1971, 1974, 1975ab, 1976] data were all gathered from the same woods behind the Institute for Advanced Study in Princeton University. He stated [1975a] that forest succession in these woods was "documented thoroughly" in his publication in American Scientist [Horn 1975b]. Horn's use of the Markov model was based, as he noted, on several assumptions:

1. Abundance of a sapling under a canopy is a reasonable predictor of survival to reach the canopy.
2. These "transition probabilities" do not change with forest composition.
3. The transition probabilities do not change with successional stage or with local edaphic conditions.

Horn [1976] commented "If recruitment of young plants is generally proportional to the local abundance of conspecific adults, the consequences for successional theory are profound." They would be profound indeed, especially since it was recognized long ago that young tree production most commonly is not proportional to conspecific adults and Horn offers no evidence to the contrary. Horn also observed that saplings can grow in many places where adult trees cannot. Oddly, he does not comment on the equally profound and familiar observation that adults of many species grow very well in places where their saplings do not grow. Horn is a recent addition to the long tradition which recognizes that age structure and reproductive classes give clues to the future trends of a forest [Daubenmire 1968]. As he commented, "linear models have made gratifying predictions of what is already known from earlier crude observations of age and reproduction distributions in forests." It comes then as no surprise that gray birch or aspen are replaced in his model by red maple or beech. Horn's promise of "rigorous analysis" however, still faces the "challenge," [1975a] of a biologically realistic measurement of the transition matrix, which is the essence of his model, based on demonstration of the validity of the assumption that the presence of a sapling under an adult tree is a basis for a precise estimate of the probabilities of replacement of that adult, and that these replacement probabilities do not change with changes in site conditions or composition. He has made little apparent progress in meeting this challenge in the several years since he first enunciated his ideas of succession as a statistical process. Like Levin's alpha values, the interest of a Markov model lies in the validity of the numbers to be plugged in and the applicability of the Markov recipe to the biological phenomenon, not solely in its mathematical rigor.

Horn is inordinately fond of distinguishing the trivial from the profound. What may be as trivial as anything he has discovered in the writings of traditional ecologists is that if one could measure the probabilities of each individual adult tree being replaced in situ by one other tree of the same or another species the "patterns of succession are direct consequences of stochastic replacements of one plant by another." Horn's assertion that the process of succession is a statistical result of "direct consequences of ergodic theorems for Markov process" should not be taken as an indication that he entirely neglects biological considerations. He provided [Horn 1971] the mechanism, competition, and in large part the limiting resource, light, which underlies his concept of tree geometry and his theory of forest structure and succession. It is apparent that the biological properties of the plants, especially light tolerance, are an important basis for the transition probabilities. Yet one finds odd juxtapositions in Horn's writings as if his mathematical rigor and logic gets in the way of his biological intuition. Horn [1974] wrote, of succession, "These patterns are independent of biological adaptations to differential successional status, though adaptations affect the speed and clarity of the patterns." The next sentence begins "Appropriate adaptations among early successional species include . . ."

Horn [1974], like Drury and Nisbet, promised "new results" which appear in Horn [1975a]. The crux of anything that may be new is the validity of the assumptions of his model that the presence of a sapling under an adult tree is a basis for estimating the probability of replacing it and that these transition probabilities do not change with change in composition, time or site conditions. These are questionable assumptions [Austin 1977]. Auclair and Cottam [1971], for example, reviewed the dynamics of deciduous oak forest in Wisconsin with particular reference to black cherry. They commented that, "black cherry accounted for approximately one-half of the total numbers in small tree and sapling data. In some cases it was the only tree species in the understory. The shade intolerance of black cherry and lack of evidence that it successfully replaces the oaks directed attention to the probable future of the species." The replacement probabilities of this and other species, in what is probably the best documented forest area in the United States, are much less clear than Horn suggested. They are not proportionate to presence of saplings under adults, but vary on different site conditions, and are influenced by changes in composition [cf. Zedler and Goff 1975]. It is very likely that Horn's probability of the beech replacing itself would change if sugar maple were present in the stands he used to calculate his probabilities of replacement. Forcier [1975] recorded a positive association of beech saplings with sugar maple canopy trees and a negative relation with beech canopy trees. Sugar maple saplings were not significantly associated with canopy individuals of either species. Whittaker and Levins [1977], citing data from Smith, reported that sugar maple saplings were positively associated with beech canopy trees and negatively with sugar maple trees, while beech in this case did not show much difference. McIntosh [1972], in studies of Catskill Mountain forests, showed that in mixed stands of beech and sugar maple there is a higher ratio of beech saplings to sugar maple saplings than in canopy trees. Size-class distributions suggested that beech and sugar maple are inverse to each other. Beech may replace itself if sugar maple is not present, but not if it is, root sprouts notwithstanding.

The problem is, as Horn [1976] comments in his most recent article on succession, that "The only sweeping generalization that can safely be made about succession is that it shows a bewildering variety of patterns." This is true in his experience even after it has carefully been restricted mostly to secondary succession in a limited forest area. Horn is learning the same lesson that traditional plant ecologists learned previously, namely that succession can be confusing. In 1975 he wrote, "The most dramatic property of succession is its repeatable convergence on the same climax community from any of many different starting points" [Horn 1975b]. In 1976 he wrote, "Succession may lead to alternative stationary states, depending on the initial composition." Horn's [1976] statement, "Analytical models dispel some of the bewilderment by showing that the general pattern of succession is largely determined by biologically interpretable properties of individual species that take part in the succession" may be questioned. He and other ecologists are still bewildered. Horn's question, "Whether succession is convergent or not depends critically on how strongly the amount of recruitment of young plants is determined by the local density of mature plants of the same species?" is an important one. Auclair and Cottam [1971], echoing many earlier ecologists, had already answered it succinctly for a large class of forests. "There is an inverse relation between the importance of trees and saplings of a species." Horn's basic data set is essentially a measure of association of adult trees and saplings, and his transition matrix is based on the probabilities of saplings being associated with canopies of adults. Zedler and Goff [1973] analyzed successional trends in another Wisconsin forest region, using an elaboration of association analysis for size class distributions, and noted an important effect of the size of the sample area on the apparent relations. They said that red maple reproduced abundantly under white and red pine, giving the impression that red maple would replace the pines. In fact, they commented, this almost never happens. Red maple is, of course, a notoriously variable species and may be expected to behave differently in different places. However, the studies of Auclair and Cottam and Zedler and Goff are a substantial step toward the requirement lately seen by Horn [1976]. "The relationship should be explored for a complex community."

Schaffer and Leigh [1976] considered the limitations of mathematical theory in plant ecology and found Horn's assumptions justified. They noted a gross overestimate of red maple in the climax, a peculiarity that Horn's model shares with Leak's [1970] model of forest succession in the northeastern United States. Schaffer and Leigh commented that Horn's model depends on the constancy of the association values which they suggest depend on the "peculiar heterogeneity of the many forest stands that make up the Institute Woods." The point made by Zedler and Goff that red maple, although abundant under pines, almost never replaced them, suggests the inadequacy of both Horn's Markovian model and Leak's birth and death model both of which predicted large future populations of red maple in mixed hardwood or beech forests.

Pickett [1976] joined Drury, Nisbet and Horn in noting the demise of the Clementsian concept of succession and offered "an evolutionary model of succession which may form the basis of a contemporary model." Like many traditional plant and contemporary animal ecologists, Pickett followed Horn in attributing major importance to competition, past or present, in arranging species in spatial and chronological gradients. He noted the absence of a contemporary model of succession saying that it is "informally approached as a collection of trends, many of which are not parallel or strictly directional." Pickett reviewed the literature which illustrates the effect of "abiotic and competitive selection pressures" in producing individualistic species and subspecific responses to gradients. He then developed the analogy between spatial and temporal (seral) gradients and the analogous causes of species distribution on these gradients. He used "analogy" rather than the more restricted "observed congruence" between spatial and temporal sequences used by Drury and Nisbet [1973].

Pickett developed an evolutionary population-based interpretation of succession specifying the following points:

1. A population cannot be both generalist and a specialist. He commented that selection preserves adaptation to niche and "fosters population coupling" [a phrase that would be grist to Senator Proxmire's mill if it ever appeared in the title of a National Science Foundation award].
2. Succession provides a complex gradient of physical and biotic environments analogous to spatial gradients.
3. Evolutionary strategy [sic] and life cycle characteristics determine the position of a species in a successional gradient.
4. Patches of different successional environments are continuously changing, depending on disturbances, offering different opportunities and selective pressures.

That species have evolved in response to different evolutionary forces in communities is a position with which few ecologists would take exception. Cowles [1904], the American pioneer of the succession, wrote, "If ecology has a place in modern biology, certainly one of its great tasks is to unravel the mysteries of adaptation," and Ganong [1904] provided seven cardinal principles of ecology all concerned with adaptation. A contemporary evolutionary geneticist, Lewontin [1969], saw Clements' theory of succession as "nothing if not an evolutionary theory of the community." Not all that much has been accomplished to link ecology and evolution, but certainly the current effort is now well publicized. Evolutionary strategies, niche and r and k selection are very much in vogue. Harper [1977] commented, "I doubt that we will be able to understand the evolutionary forces operating within plant communities until we have faced the issue of determining the behavior of genets within communities." This admonition to address the intraspecific variation in time and space greatly compounds the difficulties of ecologists dealing with more than a very limited community. A major, continuing problem lies in determining the effect of selection pressures and population processes in a context of community organization. Classical community and succession theory made it easy as groups of species were assumed to occur together in consistent associations and uniform habitats; thus, the selection pressures were generally such as to "foster population coupling." Alternative concepts of community organization [McIntosh 1967; Whittaker 1967] and divergent views of the forces structuring communities [Weins 1977; Diamond 1978] make the problem of bringing evolutionary, population and community theory together much more difficult.

Pickett's summary point is that "Successional gradients and the evolutionary and functional responses of populations on them are part of a dynamic regional process rather than a single site pattern." It is perfectly reasonable and familiar that different sites may be in different seral stages ranging from pioneer to climax, but to divorce succession from what happens on a single site over time surely is a modification of the entire concept. Any site is manifestly influenced by proximity to seed sources and other organisms that may impinge on it from adjacent sites; it is assuredly part of a landscape pattern. Succession, however, occurs on a site over time and, as noted above, a major failing of succession studies has been seen in their failure to follow the chronological sequence on a single site. Pickett's interpretation of landscape pattern and selection patterns is subject to a number of questions. The relative availability of pioneer sites versus climax sites no doubt varies greatly; in some cases, the incidence of fire for example, fairly accurate estimates may be made of availability of disturbed sites. However, one may doubt his assertion that throughout the history of an area pioneer habitats are numerically common, with diversity limited by environmental severity, and climax habitats less common. Whittaker [1976] remarks on the extreme diversity of the vegetation of areas in the Mideast which are extremely disturbed and severe.

Pickett accepted the assertions of Drury and Nisbet [1971, 1973] concerning the relation of succession of land form and classical succession and developed a three-dimensional representation of a regional vegetational surface (Figure 4) in accord with "modern concepts of geomorphology." His regional vegetation surface showed the response surface of a "succession index" plotted vertically on the plane coordinates of each site. He suggested a number of possible indices with the proviso that the index must be strictly monotonic. Possible candidates were species characteristics, the ratio between system production and respiration, the weighted average of shade adaptation (light tolerance) of species or the average degree of genetic recombination. He saw a coincidence (with slight lags) of biotic and physical parameters. Pickett started his article stating that "the classical interpretation of succession as development of vegetation through discrete stages culminating in a regional climax has been abandoned by modern ecologists." However, his regional vegetational surface was seen as ranging from a low plane surface, or entirely pioneer area, to a high, entirely climax, plane surface with

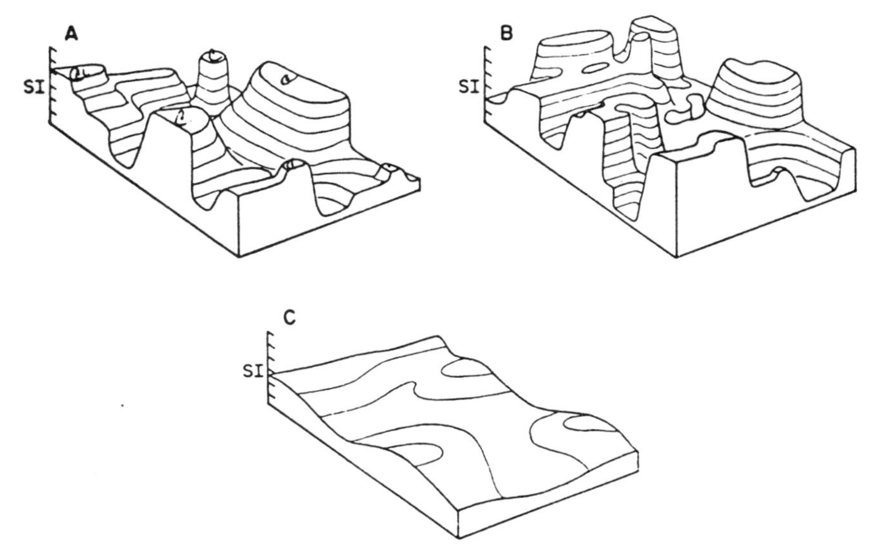

Figure 4. A regional vegetation surface. The x and y plane is a regional map. A monotonic index of successional advancement. SI, is plotted on the z axis. Changes in the configuration of the surface through time, for example the sequence A–C, illustrates the successional status and relationships throughout the region. The degree of convolution shown by the surface provides an index of γ-diversity. The region is less diverse in state C than in A. Species migrate between regions having similar SIs. Source: S. I. A. Pickett, "Succession and evolutionary interpretation," *American Naturalist* 110:107–119 (1976).

y diversity = 1, which sounds like a monoclimax. The assumption of a common montonic index linking all sites with a putative climax does not coincide easily with "Succession can be understood solely in terms of the interaction of evolutionary strategies without reference to a deterministic progression toward climax." The regional vegetational diagram looks very much like a dissected peneplain.

The most recent evaluation of succession to gain much attention is that of Connell and Slatyer [1977]. They attempt a codification of models of succession, review evidence for these, suggest predictions and tests for their models and discuss stability and community organization. They consider only the traditional concept of succession as change in species composition, which some ecosystems ecologists regard as irrelevant [O'Neill 1976]. Connell and Slatyer, like all their recent predecessors, attack Clements' theory of succession and climax and indicate that it was so satisfying to ecologists that it has dominated the field ever since, citing Odum [1969] as its perpetuator. They note the earlier questions raised by Gleason, but comment that the "queries and objections have recently increased in number." They do not cite the early objections of Tansley or the reservations of Cooper; no mention is made of the most devasting evaluations of Clementsian succession and climax theory of Whittaker [1951, 1953]. In fact, by the time of the "recent increase" in "queries and objections" in the 1960s and 1970s, most of the substance of Clementsian theory (e.g., association, monoclimax, progressive succession and organism) were substantially passé among most plant ecologists; although individuals, e.g., Daubenmire [1968] still carried on substantial elements of Clementsian concepts. Its persistence seems largely to be in textbooks, where its air of universality and orderliness makes for pedagogical convenience. Drury and Nisbet [1973] even found it necessary to attack organismic succession as perpetuated in two high school textbooks. It may also persist because most recent reassessments of succession berate the monolithic stereotype of succession raised by Clements, since it is much more difficult to attack the highly diverse and elusive concepts of succession which are actually held by most ecologists. The major elements of an organismic view of succession are perpetuated in ecosystem ecology in the writings of Margalef and Odum. These authors retain much of the Clementsian concept but shift the emphasis from compositional changes to a regional climax to collective ecosystem attributes such as nutrient or energy flow in the development of unspecified mature ecosystems.

Connell and Slatyer consider the mechanisms of succession noting, first, in contrast to Drury and Nisbet's "observed convergence," the difficulty of reconstructing chronological sequences from spatial sequences. They wrote:

> Second some possible mechanisms have been *ignored,* (italics mine) particularly the effects of grazing animals. The study of succession has in the past been carried out mainly by persons working solely with plants. . . . However, it has meant that the mechanisms concerned have usually been restricted to the interaction of plants with the physical environment or with other plants. The interactions with organisms that consume plants have always been included as one of the many factors influencing succession, but again most of the attention has been given to the consumers involved in the cycling of mineral nutrients, particularly the decomposers such as microorganisms and fungi rather than to animal herbivores.

It is certainly true that succession was mainly developed by plant ecologists, but it is an excessive statement that the effects of herbivores were "ignored." Tansley [1935] wrote that biotic factors such as heavy and continuous grazing, which changes and stabilizes the vegetation, could be a decisive influence. Tansley recognized, for example, the conversion of forest or heath to grassland. Clements coined the word "disclimax" which included, most importantly, deviations from his "true" climatic climax created by the influence of man or animals. Clements recognized extensive areas, rightly or wrongly, as deflected from the "true" pathway of succession, not only by domestic animals but also by prairie dogs and kangaroo rats. He even had a name "therium," fortunately long since discarded, for the latter. Among several early plant ecologists, Sampson [1919] produced extensive studies of grazing effects on western grasslands. In the 1930s and 1940s other range ecologists such as Costello [1944] and Dyksterhuis [1948] studied the effects of grazing on succession. Costello, for example, noted the effects of kangaroo rats, jack rabbits and harvester ants, the latter in affecting seed supply and providing by their mounds breaks in the vegetation, allowing shortgrasses to enter the succession. Dyksterhius [1948] noted that grazing could arrest a sere and developed the idea of "increaser" and "decreaser" plant species as a measure of the response of vegetation to varying degrees of grazing which was also part of the conventional wisdom of plant ecologists. Other ecologists [e.g., Nelson 1938; Smith 1940; Piemeisel 1945; Ellison 1960; Penfound 1964] recorded effects of both domestic and nondomestic animals on grassland succession.

It might be said that forest ecologists were less emphatic about the role of animals in succession than grassland ecologists, but even here the effect of animals was familiar. Minckler [1946] noted the effect of rodents on tree mortality in old-field succession. Stearns [1946] found that palatable species, such as American yew, largely disappeared from the northern hardwood forest due to deer browsing and certainly plant ecologists recognized the aptness of Leopold's "plimsoll line" as describing the effect on a forest of intense deer browing. It was long a favorite gambit of plant ecologists to frustrate ecological neophytes by asking them to identify witch hobble *(Viburnum alnifolium),* which looks very different when subjected to deer browse as it frequently was in the 1940s and 1950s. It is heartening to find Connell [1975] stating that much of the experimental evidence for predation as an important factor in community structure is derived from studies of predation on plants by mammals and citing a number of plant ecologists.

The point here is not that plant ecologists knew all about grazing effects or that they had modern quantitative appreciation or studies of the population effects of grazing on plants, but simply to record that they knew a lot more than is implied in the statement that they ignored the effects of grazing. Plant ecologists knew that species were dispersed by grazing and seed eating animals; populations could be sharply reduced or increased under the influence of grazing and browsing; plants differed widely in palatability; and successions could be modified or controlled by native or domestic animals. They were certainly aware of the effects of larger mammalian herbivores, rodents and even a less obvious activity such as mymechory was described in 1906 by Sernander, a plant ecologist. The standard techniques of studying grazing on vegetation, exclosure and enclosure, were described by Clements in his early publications and used by plant ecologists and range managers to illustrate the gross changes in composition of vegetation affected by grazing–short, perhaps, on statistical tests and sophisticated experimental design but hardly ignored.

The irony of the incorrect assumption that plant ecologists ignored grazing is seen in Connell and Slatyer's own procedure:

> We will direct our attention here to the succession of species that occupy the surface and modify the local physical conditions, e.g., plants and sessile aquatic animals. Other organisms such as herbivores, predators, pathogens, 'etc' will be included only when they affect the distribution and abundance of the main occupiers of space.

This is what many traditional plant ecologists did, and for the same reason. It is extremely difficult to encompass all aspects of succession as Connell and Slatyer's exclusions suggest. Except for Shelford, early animal ecologists were generally content to leave succession and the associated concept of community to the plant ecologists. It can be argued, contrary to Connell and Slatyer, that attention of plant ecologists was mostly given to larger animals and all too little to decomposers, such as bacteria, fungi and nutrient cycling, by plant ecologists or anyone else.

The presumed result of ignoring grazing, had that been true, was to focus attention on plant resources and the role of competition in the structure of communities. This did not, as Connell and Slatyer assert, coincide "with the development of a theory of community structure based almost entirely on competition," developed by animal ecologists with a theoretical bent in the late 1950s and 1960s, nor is it newly designated in "recent critical reviews of ecological succession." The belief that competition is a major factor in community organization has, rightly or wrongly, long been explicit in the writings of plant ecologists [Tansley 1917; Clements 1928; McIntosh 1970]. As in many current papers speculating about the role of competition in community structure, actual demonstration of competition was usually lacking, although Tansley and Clements did some of the early experimental studies of competition and Clements wrote the first treatise on the subject. It is quite true that many plant ecologists, like a major "invisible college" of theoretical animal ecologists heavily influenced by MacArthur's work, considered competition a major force in plant communities; but this is a long tradition from the earliest days and extends into recent work of Whittaker [1975]. The illusion that plant ecologists ignored competition is perpetuated by Diamond [1978]. He writes "Even today, plant ecologists tend to stress physical limiting factors almost to the exclusion of interspecific competition," a view which could not be more unrepresentative. Plant ecologists must share with MacArthur and other animal ecologists the credit or onus of asserting the primacy of competition in structuring the community [cf. Weins 1977; Pyke et al. 1977].

One of the difficulties in dealing with the problem of succession is confounding primary and secondary succession. Most of the recent papers analyzing succession have restricted their attention to secondary succession. Connell and Slatyer's paper illustrates this difficulty of dealing with the overall problem of succession in that their commentary and the first step of their proposed series of models restricts their consideration to secondary succession, although they subsequently allude to primary succession and presumably are considering it. They state: "Succession, as represented by steps A through F in figure 1 is the process by which a community recovers from a perturbation." Primary succession involves succession on sites where there is no preexisting community to be perturbed or at least nothing is left of an original community to recover. By definition it is occupation of a site uninfluenced by prior occupancy; it is familiar that prior occupancy has very marked influence on secondary succession.

Connell and Slatyer concentrate their attention on mechanisms of succession which they assert have not been defined clearly or stated as hypotheses subject to field experiment. In a rather inauspicious beginning on definition, they cite Tansley's [1935] paper as equating autogenic succession with absence of abiotic change in the physical environment. Tansley defined autogenic succession as changes induced by the *vegetation,* allogenic succession being anything else. This is the way Egler [1954] uses it, animals being an allogenic factor. Since animal ecologists have taken a hand in succession, autogenic has become synonymous with biotic [cf. Odum 1971]. Connell and Slatyer use autogenic in this sense as the mechanisms which bring about changes in succession. Connell and Slatyer review a number of instances of succession in the context of their proposed model system. It includes a number of interpretations of doubtful validity. For example, although they state that they restrict their considerations to autogenic changes in the absence of significant changes in the physical regime, they cite two examples of changes on a flood plain involving alluvial deposits by floods which are clearly not autogenic as defined by Tansley or as amended by Odum.

Their review of models of succession could be a useful codification of ideas well established in the literature, but the models are not entirely clearly stated nor are the biological examples, to the point in some instances. Connell and Slatyer attribute their model 1 (facilitation model) to relay floristics and models 2 and 3 to initial floristics, which, they state, Egler was the first to distinguish. Egler's reputation as a productive ecologist is safely established on many grounds; it will not diminish it to assert that he provided an apt term for the defining quality of secondary succession, stated by Clements, and laid more stress on it than most earlier ecologists. The point of Egler's initial floristics is that the organisms are not "arriving species." The initial flora is already present at abandonment as either or both seeds and vegetative states. Egler [1954] said, "After abandonment development unfolds from this initial flora, *without additional increments by further invasion* (for the purpose of this discussion)" (italics mine). Connell and Slatyer state that all their models agree on colonizing characteristics of species but differ in the mechanisms that determine subsequent appearance in the sequence. It does not seem likely that models 2 and 3 place the same demand on high seed production, dispersability and ability to tolerate extreme sites as does model 1. Model 1 places a premium on these qualities since the species must seize upon a new site, usually extreme, after it appears; models 2 and 3, if based on initial floristics, produce plants from material already there on a site already modified by organisms. If anything, the premium here is on viability and rapid growth as the seeds could have arrived at their leisure prior to the disturbance which created the opening. They subsequently "predict" that their facilitation model 1 will apply to situations "in which the substrate has not been influenced by organisms before hand." This is by definition a primary succession. This "prediction" has been the basis of interpretation of primary succession from Clements on and experimentally demonstrated by McCormick's work cited below. They follow with comments on secondary succession predicting that models 2 and 3 will apply there. However, as they note, "If the previous occupation has not influenced the substrate (e.g., on marine rock surfaces), however, model 1 may apply." They have, of course, defined a primary succession, so according to their prediction it should. The statement that the substrate has not been influenced before by organisms must be interpreted carefully. Manifestly, the difference between a primary and a secondary succession is that the latter has a residual effect of prior occupation even if it is laid bare by disturbance. The degree of residual effect and therefore the influence of new occupants on the site is not equivalent in all instances of secondary succession. It is always difficult to fit all organisms into neat categories. In models 2 and 3, for example, Connell and Slatyer state the pioneer species are killed by competition of later plants. Wilson and Rice [1968] and McCormick [1968] report pioneer weeds which commit suicide, presumably by an autotoxic effect, and are unable to replace themselves even in the absence of competitors; allelopathy may have other anomalous affects on succession [Rice 1976].

Connell and Slatyer review the evidence for their models citing some observations of primary successions which support the traditional model of primary succession as mediated by the effects of early species modifying the environment making it less habitable for themselves but more habitable for later arrivals. They find only one terrestrial example of experimental work supporting this view. They ignore the series of studies by McCormick and his students on the development of small terrestrial ecosystems on granite outcrops (Sharitz and McCormick 1973; McCormick et al. 1975; McCormick in press). In this elegant combination of observation and experimental studies of primary xeric succession, they showed the succession of three herbaceous species on bare rock and the influences of increasing soil depth, nutrient supply, water availability, temperature and competition. The zonal sequence in this case is demonstrated to parallel the chronological sequence and is related to the physiological tolerance and competitive abilities of the plants. These pioneer species are restricted on the extreme side of their habitat by abiotic factors and on the more moderate side by competition of other species which perform better in the deeper soils and better nutrient and moisture conditions associated with the expansion of the plant cover over the granite substrate and the weathering of the rock. McCormick commented that the implications of population interactions may be most visible during primary succession and in the pioneer stages.

Connell and Slatyer cite as evidence in intermediate stages the work of Toumey and Kienholz [1931] and Korstian and Coile [1938] as showing that even in the high light levels of early succession forests, the late succession seedlings are suppressed by root competition. They generalize that these experiments indicate that "high tolerance of later successional species to low levels of resources still does not allow them to grow to maturity if they are dominated by a stand of early species." The burden of the trenching experiments was an effort to show the importance of soil moisture as affected by root competition versus light not that later successional species cannot grow in a stand of earlier species. Studies by Kramer and Decker [1944] showed *why* hardwood seedlings thrived under pine and hardwood stands. Lutz [1945] found that hemlock *had* grown vigorously in the Toumey-Kienholz plots, although pine had died 21 years after the trenching experiments. They concluded that the shade-tolerant hemlock could grow in the pine forest. Oosting and Kramer [1946] reported similar effects in the Korstian-Coile plots; hardwoods were increasing while the pines were declining. The full range of observations in these experiments does not support the conclusions of Connell and Slatyer. In fact, the entire history of ecology in the eastern forest has shown that the early succession pine forests are replaced by later seral species such as hemlock or various hardwoods except on very extreme sites or where disturbance by fire or grazing intervenes. The growth of more tolerant species may be slowed by the presence of earlier species but certainly they do more than simply survive [Curtis 1959; Quarterman and Keever 1962].

Connell and Slatyer's interpretation of succession reminds one that two persons may report of the same water glass that it is half empty or that it is half full. They cite Henry and Swan's [1974] finding that white pines following disturbance dominate the forest for 200–250 years, "suppressing almost all later tree invasion." An old growth white pine forest will commonly have a substantial understory of hardwoods [Curtis 1959], so it is clear that the pines are a one-generation stand. One generation is 250 years; a long time but still only one generation, and successful invasions commonly take place, which is why pine forests did not replace those which were cut. The same generalization may be made of pine forests in the southeast [Quarterman and Keever 1962]. As Connell and Slatyer note, the observation time should be at least as long as the longest generation time of any of the species and any observation period approximating the life span of pines will show that they are being replaced.

Perhaps the major distinctive emphasis of the paper by Connell and Slatyer is that they address the problem of succession in marine aquatic communities; they also emphasize the importance of animals as herbivores (predators) and of pathogens. Although they incorporate all biotic elements they do not endorse the organismic view of the ecosystem and emphasize the population-centered approach to succession.

HOLISM VS INDIVIDUALISM

The common ground of the several recent reviews and commentaries on successions is:

1. They are explicitly critical of Clements' organismic, holistic concepts of succession and, implicitly or explicitly, of current organismic and holistic views as expressed by Margalef and Odum.
2. They advocate a Gleasonian, individualistic, population-centered approach to succession in contrast to an ecosystem approach.

This may be seen as a simple dichotomy based on a different choice of vantage point from which to view the ecosystem [Levin 1976]. It may, however, be seen as a more fundamental difference in level of maturity within a scientific discipline. Rickleffs [1977] sees ecologists shifting from a Clementsian holistic view of nature toward a Gleasonian or analytical appraisal of nature as a sequence of developmental stages (i.e., maturing). Clearly, not all ecologists have matured in the same direction or at the same rate.

The distinction between the population centered or individualistic position and the ecosystem or organismic position is commonly seen as a dichotomy between reductionist and holistic approaches, although some do not see the sharp distinction [Rosenzweig 1976]. The statements of ecologists may express a tolerant, live-and-let-live attitude [Levin 1976]; but there is generally little doubt where their hearts lie. Lane et al. [1974] wrote in their prospectus that, "a blend of holism and reductionism should be achieved," but in the body of their paper they were less conciliatory:

> "In ecology, there has been too much reliance on the assumption that once small portions of the system are studied, the whole can be reconstituted from the parts. There is little theoretical or experimental evidence to support this assumption."

Levins [1968b] was equally explicit:

> "Therefore an adequate science on which to base our ecological technology must be holistic, focusing on systems properties of populations and communities."

Odum [1977], like Lane et al. [1974], calls for a combination of holism with reductionism. He comments that a handful of systems ecologists managed to link together some of the reductionists in the Grassland Biome by providing something approaching an ecosystem-level model and thereby almost salvaged the International Biological Program. He sees the "new ecology" as a new integrative discipline that "deals with the supra-individual levels of organization." According to Odum, observations and experiments with natural ecological succession led to a theory that "new ecosystems properties emerge in the course of ecological development and that it is these properties that largely account for the species and growth form changes that occur." He recognizes and says he welcomes the alternative theory based on species-level processes advanced by Drury, Nisbet, Horn, Pickett, Connell and Slatyer. Others are dubious about the merit of that theory. O'Neill [1976] wrote, "There is no a priori reason to believe that the explanation for ecosystem phenomena will be found by examining populations."

Some population ecologists, however, clearly see the role of population-based, evolutionary ecology as providing the basis for understanding of ecosystem phenomena. Foin and Jain [1977], in contrast to Odum [1977], see the lack of a general, ecosystem-level theory as a requirement for "detailed work on populations because the results promise to be more lasting." They view ecosystem description as simply a preliminary to population studies because "understanding of community level processes, at least in terrestrial plant communities, will largely require solutions to life history tactics in natural communities based on researches on life histories, ecotypic variation, genetic regulation and species interaction." In their view, "holistic approaches and mathematical modeling do not inevitably overcome the traditional barriers to increasing scientific knowledge."

Harper [1977] sees ecology as a triangular science, the vertices being population ecology, production (essentially ecosystem) ecology and community ecology. Harper sees a close parallel between Gleason's individualistic concept of vegetation and the concepts of animal populations, as seen by the animal ecologists, Andrewartha and Birch, which he says are based on historical events. He asserts that the contrasting Clementsian view of community is constrained by its future, an idea which he says has many similarities with the views of the animal ecologists, Nicholson and Varley. The latter, or holistic, interpretation, sees the community as driven towards a stable state and is apparent not only in Clements "climax" but in Margalef's and Odum's "mature" state. Harper wrote, "One of the dangers of the systems approach to community productivity is that it may tempt the investigator to treat the behavior that he discovers as something that can be interpreted as if community function is organized, optimized, maximized or stabilized." He emphatically rejects any holistic idea which suggests that ecosystems are too complex for understanding by reductionists. He urges, as a model, the reductionist approach of biochemists to complex protein molecules where, "the great leaps of understanding were made by those who were willing to simplify the complexity and, as an act of faith assume that the complex whole is no more than the sum of the components and their interactions. The development of plant ecology into a predictive and rigid science depends on a similar willingness . . ." Harper suggests that the important advances in terrestrial plant ecology will come from studies of intraspecific variation within populations and intrahabitat variation. This is a population-centered view with a vengeance.

Antonovics [1976], voicing the position of the "new ecological genetics," said that the individualistic or population-based view of ecosystem is predicated on a typological view of the species. He calls for an escape from such views to thinking of communities as composed of evolving and coevolving species. Antonovics joined Harper [1977] in seeing analysis at the population level as essential to understanding "ecological and evolutionary causation." He regards the "semiphilosophical arguments about the individualistic vs organismal nature of plant communities," and presumably their ecosystem descendants, as "nonoperational." Antonovics urged, as a model, the theoretical studies of animal populations saying that plants are particularly useful for population studies, although he questioned the utility of Lotka-Volterra models.

Schaffer and Leigh [1976] note the much lamented lack of population theory in plant ecology as compared to animal ecology. They review the efforts of a number of ecologists, including Horn [1975], and argue that the theoretical population biology developed by animal ecologists is inapplicable to plants because of their heterogeneity in distribution, whereas the models of population theory assume homogeneity. Ironically, they ascribe the nonutility of such population theory to the lack of ability to describe. There may be, it appears, some utility in descriptive ecology which has frequently been downgraded [Harper 1977] or pejoratively alluded to as "mere description." Richard Lewontin [1968] voiced the call for an analytic reductionist approach to ecology and frankly deplored "biologists who reject the analytic method and insist that the problems of evolution and ecology are so complex that they cannot be treated except by holistic statements."

The differences between the descendants of the holistic and individualistic conceptualizations of nature continue in current discourse on ecology and succession, and it is useful to recognize the roots of the differences and also the continuity or lack of it [Golley 1977]. Succession to the ecosystem ecologist is a mix of continuance of the traditional concept and break with it. On one hand it continues the organismic tradition of a unitary orderly system with defined properties and predictable trends to a stable state; on the other hand it may ignore the traditional emphasis on change in composition as a major criterion of succession. According to O'Neill [1976] the ecosystem does not change when species change, "Not unless the property or measurement under study changes. It is possible for properties such as nutrient retention time to remain constant even though species change. Further, the identity of the system remains through successional changes in species." This conceptualization of an ecosystem as "fundamentally an energy processing system" whose properties "persist even though populations change" is one pole of the current views of succession. An ecosystem is seen as a definable system with measurable properties which cycles nutrients, transforms energy and produces such collective biological properties as diversity, biomass, productivity, trophic structure and stability as averages of population level processes. Species composition is incidental.

The systems concept of succession as voiced by O'Neill divorces it from compositional change and defines it in terms of change of collective ecosystem properties, e.g., nutrient retention time. In an example of recovery of a deforested northern hardwoods system following clear cutting [Likens et al. 1978], the change in species composition would be incidental to the change in nutrient flows and other ecosystem properties; once these were reestablished it would appear that the ecosystem had recovered independent of its composition. If the identity of the ecosystem remains independent of composition, ecosystem definition must be expressed in measured ecosystem properties. Is the identity seen in any single ecosystem measure or a constellation of measures? How much change in these is allowed within a recognized ecosystem? What if calcium losses are much greater than nitrogen losses and recover more slowly? Are all ecosystem properties equally important? What degree of similarity of any or all ecosystem properties suggest that two ecosystems have converged or that an ecosystem has recovered? How does one map an ecosystem? Foin and Jain [1977] cite as an example of weakening the ecosystem level approach the importance of pin cherry populations in northeastern forest succession which was seen by Likens et al. [1978] as simply a facet of ecosystem recovery.

An important aspect of much discussion of succession is the coincidence and generality of various ecosystem properties as seen in Margalef [1968] and Odum [1969]. The ecosystem presumably develops from diverse early stages with unlike ecosystem characteristics to a mature stage. Ecosystem properties change from one end of this sequence to the other, usually most rapidly in the early stages. Most of the trends are inadequately documented; the assumed parallel increase of diversity and stability has been the subject of extensive discussion and is now rejected [Goodman 1977]. Vitousek and Reiners [1975] considered ecosystem succession and nutrient retention in forest ecosystems. Contrary to earlier assumptions that as a forest ecosystem matures its ability to conserve nutrients increases, they dissociated vegetational or biotic maturity (climax) from "steady state in the ecosystem sense." According to their interpretation, nutrient losses are higher in both young and mature ecosystems than in intermediate-aged successional ecosystems. From studies on nutrient, especially nitrogen, losses in grassland ecosystems in relation to succession, Woodmansee [1978] asserts that nutrient losses are "not a function of maturing successional vegetational stages." Thus another "trend to be expected" is subject to diverse interpretation.

One of the obvious difficulties in any evaluation of succession is the lack of consistent generalizations which allow any fairly compact overview. Most of the recent commentators on succession have deliberately restricted their scope to secondary, usually old-field, succession in temperate forested areas. This is justified by Drury and Nisbet [1973] on the grounds that there are not adequate observations in nonforested areas to permit generalization. There is in fact a substantial body of succession studies on grasslands, particularly in mixed and shortgrass grasslands [Costello 1944; Ellison 1960; Penfound 1964; Haug 1970]. There are more limited observations in desert and chaparral, and an increasing body of data from arctic and tropical regions [Webb et al. 1972] and aquatic habitats based on both traditional and the "new" ecosystems ecology. The problem is that, contrary to Drury and Nisbet's [1973] assertion that the body of observations is inadequate for generalization, the inclusion of added observations seems to forestall generalization. The restriction in the recent papers on succession of the scope of consideration of a phenomenon, which is often seen as universal, is a retreat from the difficulty of transferring generalizations from secondary successions in temperate forested areas to other areas which don't fit them [Whittaker and Levin 1977]. Generalizations have been, and are still, all too easy to erect on the basis of restricted observation; but as Whittaker and Levin [1977] note, "Time has dealt unkindly with generalizations about succession," from Clements' grandiose monoclimax to the maturity seen by Margalef; the trends seen by Odum are less trendy than anticipated. It remains to be seen whether the ecosystems approach will reveal the organization which makes living relationships symphonic rather than chaotic, as Patten [1976] anticipated, or if the symphony, if found, will resemble one by Beethoven or John Cage.

One of the obvious difficulties in the current discussion of succession is that there are more entrants bringing diverse points of view into a field originally the territory of a limited number of terrestrial plant ecologists. Margalef [1958], a marine ecologist, published his early English language paper in the same journal (Journal of General Systems) in which Drury and Nisbet [1971] published their article on succession. Although both saw their concepts of succession as informed by systems, they came to opposed conclusions about the main aspect of succession. Margalef saw it as a function of an organismic entity, Drury and Nisbet urged a population-centered view. Margalef [1968] saw succession as a process of developing self organization, increasing control of the environment and accumulating energy or information, the more mature communities exploiting the less mature. He asserted that evolution cannot be understood except in the framework of ecosystems and provided a number of characteristics of selection at different successional states. He also provided a list of changes in the ecosystem indicative of its maturity. Slobodkin [1969], in a review of Margalef's book, commented that he indicated the beginnings of several "new and potentially exciting paths" producing a "feeling that sometimes the introspective conviction becomes more poetic than scientific." More to the point, Slobodkin stated that the reader is "forced to do the job that the editors should have done," a statement which could be justly applied to many of the articles and books purporting to supply new insights into ecology in the last decade. Valiela [1971], following Margalef, argued that interchange between communities at various stages is an important aspect of succession. "In a sense the more mature communities would seem to be driving the productive mechanisms of the less mature systems by exploiting the unused energy output." On this basis he questions the meaning of studies dealing with individual stages of succession as separate units.

Much of the discussion in recent reviews and reassessments of succession is confined to terrestrial examples and to limited geographical regions notably temperate forest. Generalizations concerning succession should be widely applicable in both terrestrial and aquatic habitats worldwide. Some of the more confounded discussions concerning succession derive largely from aquatic habitats (e.g., Margalef) or terrestrial areas outside of the temperate forest. Examples which seem compelling in familiar terrestrial forests are less compelling in chaparral, desert or tundra, and even less so in plankton. Succession in aquatic microcosms was compared to forest succession and large bodies of water by Odum [1969]. He saw the same basic trends in microcosms that are characteristic of terrestrial and lake successions. Odum said that succession proceeds in a microcosm in a manner not contrary to classical limnological theory which sees lakes progressing in time from less productive (oligotrophic) to more productive (eutrophic) states. Eutrophication, according to Odum, results from addition of nutrients to a lake from *outside*. In a strict sense this is not succession by his definition, which restricts it to community-controlled processes. In Odum's view, "This is equivalent to adding nutrients to the laboratory microecosystem or fertilizing a field; the system is pushed back in successional terms to a younger or 'bloom' state." Margalef [1968] also said that oligotrophic lakes are kept at a low level of maturity by nutrients flowing into them. Hutchinson [1969] noted the conventional view of lake succession from oligotrophic to eutrophic and said that a lake could return to the oligotrophic state if the nutrient influx ceases. Eutrophication, like succession, is a concept which apparently lends itself to confusion and controversy. Since eutrophication is commonly associated with increased nutrients and productivity, it is not clear how adding nutrients causes a reversal to a more oligotrophic state when, as Odum comments, a lake will revert to a more oligotrophic state when nutrient input ceases. A complicating factor in aquatic studies is that addition of fish to a fish-free pond may cause a marked increase in standing crop of algae, which is usually taken to characterize the eutrophic state. Hrbáček [1961] showed that introducing fish into a pond depleted the larger zooplankton, which limited the grazing pressure on algae, which increased to a bloom characteristic of a eutrophic state. Chemical analyses showed a similar nutrient content in both oligotrophic and eutrophic ponds. It is difficult to relate these observations to Odum's interpretations of lake succession and to draw parallels with forest succession.

Some aquatic ecologists have disputed the applicability of Gleasons' individualistic concept to aquatic systems on the grounds that it is only applicable to terrestrial organisms. Lane [1978] accused Markarewicz and Likens [1975] of forcing the results of their study of a zooplankton community to fit Gleason's individualistic concept of natural communities. Lane comments, "much of the support for the individualistic concept has come from studies of terrestrial plant communities" and argues that the concept and terminology (importance values, community continuum) associated with it are not relevant to zooplankton communities. Lane states, "there is no particular reason why 1-mm aquatic animals should behave like 10-m trees, nor do carapaces possess the same chemical composition as bark." This is patently in error, as the individualistic concept has been supported by numerous studies of benthic and marine organisms and bird studies, as well as by Makarewicz and Likens. Lane asserts that "macroscopic properties" or "consistent indices" "are characteristic of a type of community" and "vary among different types of communities," and uses Levin's [1968] equations, which measure only "niche overlap" or mutual occurrence, to imply species integration. As Wiens [1977] points out this use is untenable, and it is not clear how Lane recognizes a community type and distinguishes between types. The criterion that the "measures have some intuitive meaning for the investigator" smacks of the traditional intuitive "soziologischer Blick" of some European phytosociologists. However, the major point is that Lane's position rests on the concept of characteristic properties of an integrated community type, and attacks the individualistic concept on the basis of Levins' [1968a,b] holistic properties of communities.

Another set of aquatic observations which is difficult to fit into the concept of orderly succession was provided by Walker [1970]. Walker identified

12 stages in an aquatic succession. He determined from a core the frequencies of transition from one stage to another in a vertical (time) sequence. Far from following a putative sequence of a hydrarch succession, half of the transitions were not in sequence and 17% were regressive.

Discussion of succession is sometimes confused by consideration of scale or lack of it. Connell and Slatyer [1977] state their position on scale and self perpetuation: "So on the scale of generation times and over a large enough tract, if both early and late-successional stages persist despite perturbations both are stable." Pickett [1976] similarly confounded succession as a regional or landscape pattern and a single site pattern. The fact that an early successional stage persists in a landscape is not despite disturbance, as Connell and Slatyer stated, but because of it. The crux of the succession problem is that a species does or does not replace itself on a given site not that it moves successfully to a new site if disturbance makes one available. Succession, climax or stability have enough problems without confusing site stability with landscape or regional stability. Loucks [1970] wrote that succession may be seen as repetitive wave patterns initiated by random disturbances and varying intervals. He examined forest stands on nonextreme sites and suggested that succession and associated community patterns followed wave patterns interrupted by severe disturbances at various intervals. Loucks said that a forest in which changes are taking place should not be regarded as unstable but as part of a series of phenomena making up a stable (homeostatic) system capable of repeating itself whenever a disturbance occurs. Hence, the entire landscape with various stages of the succession may be stable, and various species of the sere which are isolated in the time sequence may be perpetuated. This sounds very much like Clements' original position, which was that the climax association included all of the seral stages. It shifts the stage from the community at any site to a much larger stage, the regional landscape. The whole may in some sense be looked at as a homeostatic mechanism or, as Pickett asserts, a mosaic of successional habitats generated by random or periodic disturbance. That early succession or pioneer areas must recur if certain species are to persist in a region is not in dispute, but this should not be confused with a sequence on a single site. Neither should the existence of a gradational sequence on the landscape be confused with a chronological sequence on a single site. Zedler and Goff [1973] make the point very well. They noted that if their sample area had been as large as a county, the overall mixture of tree composition would have been stable, and at this large scale all species would be "climax" in the sense of persisting in the landscape. As the size of the sample area was reduced the patterns would become more representative of within-stand, or single-site succession, as it is generally understood. It is on this scale that pioneer and climax take on meaning in the ordinary context of succession as abstruse as it may appear. To have both "climax" pioneer and climax species or communities or ecosystems clearly becomes unproductive.

The pattern of vegetation or ecosystems in a landscape has long been visualized as a patchwork of seral stages. In the Clementsian view, these were developing toward, if not reaching, a regional monoclimax with a variety of semistable communities (subclimaxes), persisting for relatively long periods. Whittaker [1951, 1953, 1974] saw the landscape pattern as a continuous array of both seral and climax communities of several types incorporating the idea that something which was climax on one site could be seral on another. Whittaker and Levin [1977] recognize the landscape pattern as a mosaic of seral and climax stages but emphasize that the pattern on the larger landscape scale is conditioned substantially by types and incidence of disturbance.

The apparent polarities of individualistic and organismic views of succession are complicated by similar polarities which are seen in the factors that control community organization and succession. Traditional plant ecology ascribed to competition a major role in controlling community organization and succession [McIntosh 1970]. A major recent invisible college of theoretical animal ecology, associated with MacArthur, similarly sees competition as the primary force dictating community structure. Cody and Diamond [1975] wrote: "It is natural selection operating through competition that makes the strategic decisions on how sets of species allocate their time and energy; the outcome of the process is the segregation of species along resource utilization axes." Relatively recent opposition to this view is seen in studies suggesting that predation, rather than competition, is the major force structuring communities [Brooks and Dodson 1965; Paine 1969; Wiens 1977; Werner 1974; Pyke et al. 1977]. Pyke et al. state without equivocation: "Thus predation, as determined by the foraging behavior of animals in a community, is the core of community structure.

The ideas of succession and climax were essentially the first theoretical statement of ecology [McIntosh 1976]. Clements framed these in a highly structured, deductive, deterministic theory which provided a central, if much criticized, basis for the development of American plant ecology and which occupied a dominant position in ecological textbooks [Egler 1951]. The search for a universal generalization concerning succession has fared badly as the theory was extended from terrestrial vegetation largely in temperate forest, to other vegetations, to aquatic systems, to animal communities and to the ecosystem. The search was complicated by the change from a relatively simple concept of change of population composition on given sites as the prime criterion to a bewildering array of added criteria accompanied by a new eschatology to replace that of some of the adherents of Clements who demonstrated the fervor noted by Tansley [1935]. The traditional and persistent belief in the unity and balance of nature and the persistence of metaphysical traditions into biological discourse, noted by Simberloff (in press), are contributors to this feeling. According to this tradition, there has to be an organizing principle to the bewildering array of things ecologists see in nature, and if we don't discern it now it is simply because we do not know enough. Some ecologists, in the tradition of Albert Einstein, cannot believe that God plays dice with the ecosystem even though some of the processes are stochastic.

It is perhaps too simple to see the current spectrum of ecologists as two camps searching for significant generalities at the species or ecosystem level of organization. Yet it is clear that some ecologists see regularities in populations [e.g., Yoda's 3/2 law; Harper and White 1974] and urge that the basis for understanding ecosystems is via an amalgamation of evolutionary genetics and population theory [Lewontin 1969; Harper 1977]. Others look to ecosystem levels for their generalities [Odum 1977].

May [1974] anticipated the "perfect crystals" of ecology; in 1976 he noted there were "many examples where the world appears chaotic and vagarious at the level of individual species but nonetheless 'constant and predictable' at the level of community organization." "In this spirit" May examined the "intriguing generalization" he attributed to Slobodkin [1962] that the efficiency with which energy is transferred from one trophic level to the next is around 10%. He found that this example of a "constant and predictable value at the level of community organization . . . follows no simple and universal rule" but instead produces "an array of such figures depending on the details of the environments and organisms involved." Nothing daunted, May [1977] returns to the same subject. He notes that many introductory ecology books contain the "grand generalization" that 10% of the energy in any one trophic level is transferred to the next level. [If they do, they should not.] However, as May commented "widely disparate figures" are "a fact which has tended to discourage theoretical activity in this area." This is not the first time, as T. H. Huxley once commented, that a beautiful theory has been killed by an ugly fact. What is not mentioned in May's comments on the 10% efficiency idea is that a symposium [Slobodkin 1970] specifically designed to examine it decided that the empirical evidence did not support it. The author of the idea of the existence of a maximum ecological efficiency thought it fitting that he should be the one to deny it. Slobodkin [1972] commented, "We therefore have no reason to believe that ecological efficiency is in fact constant, and as a matter of fact it is not constant." More generally he said that any theory "containing extensive variables as a necessary component, is to be regarded with suspicion."

There is an anomaly in the fact that the traditional problem of the descriptive plant ecologist, homogeneity, or its converse heterogeneity, is now seen as critical by proponents of opposed points of view. Harper's [1977] recognition of the importance of pattern in Hopkins poetic metaphor "dappled things" is expanded in Whittaker's recognition that the dappling is not static but is, in his metaphor a "shimmer of populations." Schaffer and Leigh [1976] see the pattern inherent in vegetation as limiting the application of mathematical theory derived from animal populations, which Harper sees as a model for ecology.

The problem created by the irregular and changing nature of patterns of ecological phenomena prompted Schaffer and Leigh [1976] to argue that theoretical mathematical population ecology, developed largely by animal ecologists, is not applicable to plants since those theories presume "uniform mixing." It may be questioned on the same ground that these theories are applicable to animals which are not uniformly mixed either. Schaffer and Leigh comment that the relevance of mathematical theory to ecology hinges ironically on the mathematicians ability to describe, a distinctly old fashioned virtue. The theoreticians task they say is, "to describe spatial heterogeneity in terms simple enough to understand and yet complete enough to predict accurately *how the spatial pattern will develop and change with the passage of time*" (italics added). This problem is what the vertex of descriptive community ecologists of Harper's ecological triangle have wrestled with since the early years of this century with limited success. It remains to be seen if theoreticians will do better if they choose to undertake the task laid on them by Schaffer and Leigh. It is much simpler and mathematically more tractable to assume homogeneity and equilibrium.

The problems of succession are, not surprisingly, integrated with problems of scale and heterogeneity or pattern, both in vegetation and among ecologists. Some of the disputes may be resolved by appropriate consideration of hierarchy and scale and questions of the emergence of properties at higher levels of organization. However, the prospects for rapproachment are not good if reciprocal ignorance persists between the invisible colleges of ecology.

The search for clarity if not unity in succession has daunted ecologists from the beginning. Cowles [1899] posed the difficulty noting that, "the flora of an area must be approached not as a changeless landscape feature, but rather as a panorama, never twice alike. . . . Ecology therefore is a study in dynamics." Cowles [1901] commented: "When we say there is an approach to the mesophytic forest we speak only roughly and approximately. As a matter of fact, we have a variable approaching a variable rather than a constant." Thus, the problem of assessing approach to equilibrium was seen by some early ecologists as they recognized the difficulty of hitting a moving target. The diversity and changing nature of succession was seen by Cooper [1926] in his metaphor of succession as a braided stream. Stanley Cain [1944] expressed the concern that ecology might be too complex for mathematical analysis. Egler said that ecology may not only be more complicated than we think, it may be more complicated than we can think. Golley [1977], perhaps in despair during editing a volume endeavoring to provide an overview of succession wrote, "A simple mechanistic explanation of succession is not possible. Truly there is a rich array of possible mechanisms to explain succession." Whittaker and Levin [1977] consider the problems created by the effects of mosaic phenomena in communities and succession and comment on the bearing of these on the great diversity of ecological theory. They write,

> The failure of unifying statements on succession may not only be historic but predictive. When we discuss communities beyond their most essential attributes as open systems, generality may elude us, except for the generality of diversity. Ecological theory is not precluded by, but should make realistic allowance for, the intrinsic diversity of ecological phenomena; and ecological research must often center on more analysis, interpretation, comparison, and modeling of cases than on widely applicable generalization. Ecologists have sought a theory or master plan of evolution permitting interpretation of communities, through a limited number of strongly linked and widely significant

relationships. Such a theory is naturally desired by ecologists as scientists; but the reasoning of this paper suggests that there may be no master plan except, perhaps, the evolution of such a diversity of relationships as to frustrate that desire.

Such doubts will assuredly not deter those seeking to bring regularity to ecology via information theory, thermodynamics, linear programming, linguistics, systems analysis, catastrophe theory, network theory or transcendental meditation. The contrast in ecology between Levins' noise-free alpha values, Diamond's "assembly rules," Horn's Markovian successions without intervention of biological considerations, Odum's "trends to be expected," May's "perfect crystals" and Whittaker's elusive "shimmer of population" is evident in much of the discussion about succession today whatever the terminology. The search for satisfying regularity and simplicity is traditional in science, and there is no reason to forgo that search. There is similarly no reason to pursue an illusion that simplicity will be introduced by calling disturbance "perturbation" or secondary succession, "ecosystem recovery" or pioneer "opportunist."

ACKNOWLEDGMENTS

Numerous people were kind enough to review the manuscript of this paper at various stages of preparation. My colleagues at the Division of Environmental Biology of the National Science Foundation, John Brooks, Wayne Swank and Tom Callahan lent sympathetic ears and offered sage advice. Other long-suffering friends added it to already long reading lists. I appreciate the comments of J. Frank McCormick, Peter Rich, Robert Peet, Norman Christensen, Grant Cottam, Timothy Allen, Frances James, David Morgan and Ronald Hellenthal. Students in my Ecology Seminar at Notre Dame critically reviewed specific portions of the manuscript and Susan Schwartz efficiently transformed a handwritten, cut-and-paste mess into an orderly, legible manuscript.

REFERENCES

Allee, W. C., A. E. Emerson, O. Park, T. Park and K. P. Schmidt. *Principles of Animal Ecology* (Philadelphia, PA: W. B. Saunders Co., 1949), 837 pp.

Antonovics, J. "The input from population genetics: the new ecological genetics," *Systematic Bot.* 1:233–245 (1976).

Auclair, A. N., and G. Cottam. "Dynamics of black cherry (*Prunus serotina* Ehhr.) in southern Wisconsin oak forests," *Ecol. Monogr.* 41:153–177 (1971).

Austin, M. P. "Use of ordination and other multivariate descriptive methods to study succession," *Vegetatio* 35:165–175 (1977).

Bodenheimer, F. S. "The concept of biotic organization in synecology," in *Studies in Biology and its History*, F. S. Bodenheimer, Ed. (Jerusalem: Biological Studies, 1957), pp. 75–90.

Brookhaven Symposia in Biology 1969. "Diversity and stability in ecological systems," No. 22 (Upton, NY: Brookhaven National Laboratory), 264 pp.

Brooks, J. L., and S. I. Dodson. "Predation, body size and composition of plankton," *Science* 150:28–35 (1965).

Burgess, R. L. "The Ecological Society of America. Historical data and some preliminary analyses," in *History of American Ecology*, F. N. Egerton, Ed. (New York: Arno Press, 1977), 24 pp.

Cain, A. J. "The efficacy of natural selection in wild populations," in *Changing Scenes in the Natural Sciences, 1967-1976*, C. E. Goulden, Ed. (Philadelphia: Academy of Natural Sciences, 1977), pp. 111–134.

Cain, S. A. *Foundations of Plant Geography* (New York: Harper and Brothers, 1944), 556 pp.

Chamberlain, T. C. "The method of multiple working hypotheses," *Science* 148:754–759 (1965). (Reprint of 1890 article).

Clements, F. E. *Research Methods in Ecology* (Lincoln, NB: University Publishing Co., 1905), 334 pp.

Clements, F. E. *Plant Succession: An Analysis of the Development of Vegetation*, Publ. No. 242 (Washington, DC: Carnegie Institution, 1916).

Clements, F. E. *Plant Succession and Indicators* (New York: H. W. Wilson, 1928), 453 pp.

Clements, F. E., and V. Shelford. *Bio-ecology* (New York: John Wiley and Sons, Inc., 1939), 425 pp.

Cody, M. L., and J. M. Diamond, Eds. *Ecology and the Evolution of Communities* (Cambridge, MA: Belknap Press of Harvard University Press, 1975), 545 pp.

Colinvaux, P. *Introduction to Ecology* (New York: John Wiley and Sons, Inc., 1973), 621 pp.

Connell, J. H., "Some mechanisms producing structure in natural communities: a model and evidence from field experiments," in *Ecology and Evolution of Communities*, M. L. Cody and J. M. Diamond, Eds. (Cambridge, MA: Belknap Press of Harvard University Press, 1975), pp. 460–490.

Connell, J. H., and R. O. Slatyer. "Mechanisms of succession in natural communities and their role in community stability and organization," *Am. Nat.* 111:1119–1144 (1977).

Connor, E. F., and D. Simberloff. "The assembly of species communities: Chance or competition," *Ecology* (in press).

Cook, R. E. "Review of: Rickleffs R. Ecology," *J. Ecol.* 62:966–967 (1974).

Cook, R. E. "Raymond Lindeman and the trophic-dynamic concept in ecology," *Science* 198:22–26 (1977).

Cooper, W. S. "The climax forest of Isle Royale, Lake Superior, and its development," *Bot. Gaz.* 55:1–44, 115–140, 189–235 (1913).

Cooper, W. S. "The recent ecological history of Glacier Bay, Alaska: II. The present vegetation cycle," *Ecology* 4:223–246 (1923).

Cooper, W. S. "The fundamentals of vegetational change," *Ecology* 7:391–413 (1926).

Costello, D. F. "Natural revegetation of abandoned plowed land in the mixed prairie association of northeastern Colorado," *Ecology* 25:312–326 (1944).

Cowles, H. C. "The ecological relations of the vegetation on the sand dunes of Lake Michigan," *Bot. Gaz.* 27:95–117, 167–202, 281–308, 361–391 (1899).

Cowles, H. C. "The physiographic ecology of Chicago and vicinity: a study of the origin, development, and classification of plant societies," *Bot. Gaz.* 31:73–108, 145–182 (1901).

Cowles, H. C. "The work of the year 1903 in ecology," *Science* 19:879–885 (1904).

Cowles, H. C. "The causes of vegetative cycles," *Bot. Gaz.* 51:161–183 (1911).

Craig, R. B. "Review of: Pianka, E. R., Evolutionary Ecology," *Ecology* 57:212 (1976).

Crane, D. *Invisible Colleges* (Chicago: University of Chicago Press, 1972), 213 pp.

Curtis, J. T. *The Vegetation of Wisconsin* (Madison, WI: The University of Wisconsin Press, 1959), 657 pp.

Daubenmire, R. *Plant Communities* (New York: Harper and Row, Publishers, Inc., 1968), 300 pp.

Diamond, J. M. "Assembly of species communities," in *Ecology and Evolution of Communities*, M. L. Cody and J. M. Diamond, Eds. (Cambridge, MA: Belknap Press of Harvard University Press, 1975), pp. 342–444.

Diamond, J. M. "Niche shifts and the rediscovery of interspecific competition," *Am. Scientist* 66:322–331 (1978).

Drury, W. H., and I. C. T. Nisbet. "Inter-relations between development models in geomorphology, plant ecology, and animal ecology," *General Systems* 16:57–68 (1971).

Drury, W. H., and I. C. T. Nisbet. "Succession," *J. Arnold Arboretium* 54: 331–368 (1973).

Dyksterhius, E. J. "The vegetation of western cross timbers," *Ecol. Monog.* 18:325–376 (1948).

Egerton, F. N. "Changing concepts of the balance of nature," *Quart. Rev. Biol.* 48:322–350 (1973).

Egerton, F. N. *American Plant Ecology, 1897-1917* (New York: Arno Press, 1977).

Egler, F. E. "A commentary on American plant ecology based on the textbooks of 1947-1949," *Ecology* 32:673–695 (1951).

Egler, F. E. "Vegetation science concepts I. Initial floristics composition, a factor in old-field vegetation development," *Vegetatio* 4:412–417 (1952–54).

Ellison, L. "Influence of grazing on plant succession of rangelands," *Bot. Rev.* 26:1–78 (1960).

Elton, C. *Animal Ecology* (New York: Macmillan Publishing Co., Inc., 1927), 209 pp.

Emerson, A. E. "The evolution of adaptation in population systems," in *Evolution after Darwin, Vol. I*, S. Tax, Ed. (Chicago: University of Chicago Press, 1960), pp. 307–348.

Foin, T. C., and S. K. Jain. "Ecosystem analysis and population biology: lessons for the development of community ecology," *Bioscience* 27:532–538 (1977).
Forbes, S. A. "On some interactions of organisms," *Bull. Ill. State Lab. Nat. Hist.* 1(3):3–17 (1880).
Forbes, S. A. "The food relations of the Carabidae and Coccinellidae," *Bull. Ill. State Lab. Nat. Hist.* 1(6):33–39 (1883).
Forbes, S. A. "The lake as a microcosm," *Bull. Ill. Nat. Hist. Survey* 15:537–550 (1887).
Forcier, L. K. "Reproductive strategies and the co-occurrence of climax tree species," *Science* 189:808–809 (1975).
Fretwell, S. D. "The impact of Robert MacArthur on ecology," *Ann. Rev. Ecol. Syst.* 6:1–13 (1975).
Ganong, W. F. "The cardinal principles of ecology," *Science* 19:493–498 (1904).
Ghiselin, M. T. *The Economy of Nature* (Berkeley: University of California Press, 1974), 346 pp.
Glacken, C. J. *Traces on the Rhodian Shore* (Berkeley: University of California Press, 1967), 736 pp.
Gleason, H. A. "The vegetation of the inland sand deposits of Illinois," *Bull. Ill. State Lab. Nat. Hist.* 9:21–174 (1910).
Gleason, H. A. "The structure and development of the plant association," *Bull. Torrey Bot. Club* 44:463–481 (1917).
Gleason, H. A. "The individualistic concept of the plant association," *Bull. Torrey Bot. Club* 53:1–20 (1926).
Gleason, H. A. "Further views on the succession concept," *Ecology* 8:299–326 (1927).
Golley, F. B., Ed. *Ecological Succession* (Stroudsburg, PA: Dowden, Hutchinson and Ross, Inc., 1977), 373 pp.
Goodman, D. "The theory of diversity-stability relationships in ecology," *Quart. Rev. Biol.* 50:237–266 (1975).
Griffith, B. C., and N. C. Mullins. "Coherent social groups in scientific change," *Science* 112:959–964 (1972).
Harper, J. L. "The contributions of terrestrial plant studies to the development of the theory of ecology," in *Changing Scenes in the Natural Sciences, 1776–1976*, C. E. Goulden, Ed. (Philadelphia: Academy of Natural Sciences, 1977), pp. 139–158.
Harper, J. L., and J. White. "The demography of plants," *Ann. Rev. Ecol. Syst.* 5:419–463 (1974).
Haug, P. T. "Succession on Old Fields: A Review," M.Sc. Thesis, Colorado State University, Fort Collins, CO (1970).
Henry, J. D., and J. M. A. Swan. "Reconstructing forest history from live and dead plant material—an approach to the study of forest succession in southwest New Hampshire," *Ecology* 55:772–783 (1974).
Horn, H. S. *The Adaptive Geometry of Trees* (Princeton, NJ: Princeton University Press, 1971).
Horn, H. S. "The ecology of secondary succession," *Ann. Rev. Ecol. Syst.* 5:25–37 (1974).
Horn, H. S. "Markovian processes of forest succession," in *Ecology and Evolution of Communities*, M. L. Cody and J. M. Diamond, Eds. (Cambridge, MA: Harvard University Press, 1975a), pp. 196–211.

Horn, H. S. "Forest succession," *Scientific Am.* 232:90–98 (1975b).

Horn, H. S. "Succession," in *Theoretical Ecology*, R. M. May, Ed. (Philadelphia: W. B. Saunders Co., 1976), pp. 187–190.

Hrbáček, J., M. Dvořaková, V. Kořinek and L. Procházkóva. "Demonstration of the effect of the fish stock on the species composition of zooplankton and the intensity of metabolism of the whole plankton association," *Vehr. int. Verin. theor. angew. Limnol.* 14:192–195 (1961).

Hutchinson, G. E. "Eutrophication, past and present," in *Eutrophication: Causes, Consequences, Corrections* (Washington, DC: National Academy of Sciences, 1969), pp. 17–26.

Innis, G. S. "The use of a systems approach in biological research," *Study of Agricultural Systems*, G. Dalton, Ed. (London: Applied Science Publishers, Ltd., 1975), pp. 369–391.

Johnson, H. A. "Information theory in biology after 18 years," *Science* 168:1545–1550 (1970).

Johnson, P. L. "An ecosystem paradigm for ecology," Oak Ridge Associated Universities–129. Oak Ridge, TN (1977), 20 pp.

Juhasz-Nagy, P. "Investigations concerning ecological homeostasis (EH)," *Tenth Botany Congress* (Edinburgh, Scotland: T. and A. Constable, 1964), p. 395.

Karpov, V. G. "An experimental investigation of succession in forest biogeocoenoses in the Taiga zone," *Dokl. Akad. Nauk SSR* 156:203–206 (1964).

Keever, C. "Causes of succession on old fields of the Piedmont, North Carolina," *Ecol. Monog.* 20:231–250 (1950).

Kolata, G. G. "Theoretical ecology: beginnings of a predictive science," *Science* 183:400–401 (1972).

Korstian, C. F., and T. S. Coile. "Plant competition in forest stands," Duke University School of Forest. Bull. 3 (1938), 125 pp.

Kramer, P. J., and J. P. Decker. "Relation between light intensity and rate of photosynthesis of loblolly pine and certain hardwoods," *Plant Physiol.* 19:350–358 (1944).

Kuhn, T. S. *The Structure of Scientific Revolutions* (Chicago: University of Chicago Press, 1970), 210 pp.

Lane, P. A. "Zooplankton niches and the community structure controversy," *Science* 200:458–460 (1978).

Lane, P. A., G. H. Lauff and R. Levins. "The feasibility of using a holistic approach in ecosystem analysis," in *Ecosystem Analysis and Prediction. Proceedings of the Conference on Ecosystems, July 1–5, 1974*, S. A. Levin, Ed. (Philadelphia: Society Industrial and Applied Mathematics, 1975), pp. 111–128.

Leak, W. B. "Successional change in northern hardwoods predicted by birth and death simulation," *Ecology* 51:794–801 (1970).

Levandowsky, M. "Ecological niches of sympatric phytoplankton," *Am. Nat.* 106:71–78 (1972).

Levin, S. A., Ed. *Ecological Theory and Ecosystem Models* (Indianapolis: The Institute of Ecology, 1976), 71 pp.

Levins, R. *Evolution in Changing Environments* (Princeton, NJ: Princeton University Press, 1968a), 120 pp.

Levins, R. "Ecological engineering: Theory and technology," *Quart. Rev. Biol.* 43:301–305 (1968b).

Lewontin, R. C., Ed. *Population Biology and Evolution* (Syracuse, NY: Syracuse University Press, 1968), 206 pp.

Lewontin, R. C. "The meaning of stability," in *Diversity and Stability in Ecological Systems, Brookhaven Symp. Biol. No. 22* (1969), pp. 13–24.

Likens, G. E., F. H. Bormann, R. S. Pierce and W. A. Reiners. "Recovery of a deforested ecosystem," *Science* 199:492–496 (1978).

Livingston, R. B., and M. J. Allessio. "Buried viable seed in successional field and forest stands, Harvard Forest Massachusetts," *Bull. Torrey Bot. Club* 95:58–69 (1968).

Loucks, O. "Evolution of diversity, efficiency, and community stability," *Am. Zool.* 10:17–25 (1970).

Lutz, H. J. "Vegetation on a trenched plot twenty-one years after establishment," *Ecology* 26:200–202 (1945).

MacFadyen, A. "Some thoughts on the behavior of ecologists," *J. Anim. Ecol.* 44:351–363 (1975).

Margalef, D. R. "Information theory in ecology," *Gen. Syst.* 3:36–71 (1958).

Margalef, D. R. "On certain unifying principles in ecology," *Am. Nat.* 97: 357–374 (1963).

Margalef, D. R. *Perspectives in Ecological Theory* (Chicago: University of Chicago Press, 1968), 111 pp.

Markarewicz, J. C., and G. E. Likens. "Niche analysis of a zooplankton community," *Science* 190:1000–1002 (1975).

Marks, P. L., and F. H. Bormann. "Revegetation following forest cutting: Mechanisms for return to steady-state nutrient cycling," *Science* 176: 914–915 (1972).

May, R. M. *Stability and Complexity in Model Ecosystems* (Princeton, NJ: Princeton University Press, 1973), 236 pp.

May, R. M. "Scaling in ecology," *Science* 184:1131 (1974).

May, R. M. "Patterns in multi-species communities," *Theoretical Ecology*, R. M. May, Ed. (Philadelphia: W. B. Saunders Co., 1976), pp. 142–162.

May, R. M. "Mathematical models and ecology: Past and future," in *Changing Scenes in the Natural Sciences, 1776–1976*, Special Publ. 12, C. E. Goulden, Ed. (Philadelphia: Academy of Natural Sciences, 1977), pp. 189–202.

McCormick, J. "Succession," *Via* 1:1–16 (1968).

McCormick, J. F. "Implications of population interactions in primary succession," (in press).

McCormick, J. F., A. E. Lugo and R. Sharitz. "Experimental analysis of ecosystems," in *Vegetation and Environment*, B. R. Strain and W. C. Billings, Eds. (The Hague: Dr. W. Junk B. V. Publ., 1974), pp. 151–179.

McIntosh, R. P. "The continuum concept of vegetation," *Bot. Rev.* 33:130–187 (1967).

McIntosh, R. P. "Community, competition, and adaptation," *Quart. Rev. Biol.* 45:259–280 (1970).

McIntosh, R. P. "Forests of the Catskill Mountains, New York," *Ecol. Monog.* 41:143–161 (1972).

McIntosh, R. P. "Plant ecology. 1947–1972," *Ann. Miss. Bot. Gard.* 61:132–165 (1974).

McIntosh, R. P. "H. A. Gleason–'individualistic ecologist,' 1882–1975," *Bull. Torrey Bot. Club* 102:253–273 (1975).

McIntosh, R. P. "Ecology since 1900," in *Issues and Ideas in America*, B. J. Taylor and T. J. White, Eds. (Norman, OK: University of Oklahoma Press, 1976), pp. 353–372.

Minkler, L. S. "Old field reforestation in the great Appalachian valley as related to some ecological factors," *Ecol. Monog.* 16:87–108 (1946).

Odum, E. P. "The strategy of ecosystem development," *Science* 164:262–270 (1969).

Odum, E. P. *Fundamentals of Ecology* (Philadelphia: W. B. Saunders Company, 1971), 574 pp.

Odum, E. P. "The emergence of ecology as a new integrative discipline," *Science* 195:1289–1293 (1977).

Olson, J. S. "Rates of succession and soil changes on southern Lake Michigan sand dunes," *Bot. Gaz.* 119:125–170 (1958).

O'Neill, R. "Paradigms of ecosystem analysis," in *Ecological Theory and Ecosystem Models*, S. A. Levin, Ed. (Indianapolis: The Institute of Ecology, 1976), pp. 16–19.

Oosting, H. J., and M. E. Humphries. "Buried viable seeds in a successional series of old field and forest soils," *Bull. Torrey Bot. Club* 67:253–273 (1940).

Oosting, H. J., and P. J. Kramer. "Water and light in relation to pine reproduction," *Ecology* 27:47–53 (1946).

Orwell, G. *Nineteen Eighty-four* (New York: Harcourt Brace, 1949).

Paine, R. T. "The *Pisaster-Tegula* interaction: Prey patches, predator food preference, and intertidal community structure," *Ecology* 50:950–961 (1969).

Patten, B. C., Ed. *Systems Analysis and Simulation in Ecology. Vol. I* (New York: Academic Press, Inc., 1971), 610 pp.

Patten, B. C. "Ecosystem linearization: an evolutionary design problem," *Am. Nat.* 109:529–539 (1975).

Patten, B. C. "Afterthoughts on the Cornell TIE Workshop," in *Ecological Theory and Ecosystem Models*, S. A. Levin, Ed. (Indianapolis: The Institute of Ecology, 1976), pp. 63–66.

Penfound, W. T. "The relation of grazing to plant succession in the tallgrass prairie," *J. Range Manage.* 17:256–260 (1964).

Pickett, S. T. A. "Succession: an evolutionary interpretation," *Am. Nat.* 110:107–119 (1976).

Piemeisel, R. L. "Causes affecting change and rate of change in a vegetation of annuals in Idaho," *Ecology* 32:53–72 (1951).

Pyke, G. H., H. R. Pulliam and E. L. Charnov. "Optimal foraging: A selective review of theory and tests," *Quart. Rev. Biol.* 51:137–154 (1977).

Quarterman, E., and C. Keever. "Southern mixed hardwood forest: Climax in the southeastern coastal plain: U.S.A.," *Ecol. Monog.* 31:167–185 (1962).

Raunkaier, C. *The Life Forms of Plants and Statistical Plant Geography* (Oxford: Clarendon Press, 1934), 632 pp.

Rice, E. L. "Allelopathy and grassland improvement," in *The Grasses and Grasslands of Oklahoma*, J. R. Estes and R. J. Tyrl, Eds. (Ann. Okla. Acad. Sci., No. 6, 1976), pp. 90–112.

Rickleffs, R. E. "Introductory Remarks," in *Changing Scenes in the Natural Sciences, 1776–1976*. Special Publ. 12, C. E. Goulden, Ed. (Philadelphia: Academy of Natural Sciences, 1977), pp. 135–138.

Rosenzweig, M. L. "Review of Golley, F. B., K. Petrusewicz and L. Ryskowski (eds.). Small Mammals," *Science* 912:778–779 (1976).

Sagar, G. R., and J. L. Harper. "Factors affecting the germination and early establishment of plantains (*Plantago*)," in *The Biology of Weeds*, J. L. Harper, Ed. (London: Oxford University Press, 1960), pp. 236–246.

Sampson, A. W. "Plant succession in relation to range management," USDA Bull. 741 (1919).

Schaffer, Wm. M., and E. G. Leigh. "The prospective role of mathematical theory in plant ecology," *Syst. Bot.* 1:233–245 (1976).

Schroder, C. D., and M. L. Rosenzweig. "Perturbation analysis of competition and overlap in habitat utilization between *Dipodomys ordii* and *Dipodomys merriami*," *Oecologia* 19:9–28 (1975).

Sharitz, R. R., and J. F. McCormick. "Population dynamics of two competing plant species," *Ecology* 54:723–740 (1973).

Shelford, V. E. "Ecological succession," *Biol. Bull.* 21:127–151 (1911).

Shelford, V. E. *Animal Communities in Temperate America* (Baltimore, MD: Williams and Wilkins Co., 1913), 608 pp.

Simberloff, D. "A succession of paradigms in ecology, essentialism to materialism and probabilism," *Synthese* (in press).

Slobodkin, L. B. "Energy in animal ecology," *Adv. Ecol. Res.* 1:69–101 (1962).

Slobodkin, L. B. "Pathfinding in ecology," *Science* 164:817 (1969).

Slobodkin, L. B. "Summary of the symposium," in *Marine Food Chains*, J. H. Steele, Ed. (Edinburgh, Scotland: Oliver and Boyd, 1970), pp. 337–340.

Slobodkin, L. B. "On the inconstancy of ecological efficiency and the form of ecological theories," *Trans. Conn. Acad. Arts and Sci.* 44:293–305 (1972).

Smith, C. C. "Biotic and physiographic succession on abandoned eroded farmland," *Ecol. Monog.* 10:421–484 (1940).

Stearns, F. W. "Ninety years change in a northern hardwood forest in Wisconsin," *Ecology* 30:350–358 (1949).

Swan, J. M. A., and A. M. Gill. "The origin, spread, and consolidation of a floating bog in Harvard Pond, Petersham, Massachusetts," *Ecology* 51: 829–840 (1970).

Tansley, A. G. "On competition between *Galium saxatile* L. (*G. hercynicum Weig.*) and *Galium sylvestre* (*G. asperum* schreb.) on different types of soil," *J. Ecol.* 5:173–179 (1917).

Tansley, A. G. "The classification of vegetation and the concept of development," *J. Ecol.* 8:118–149 (1920).

Tansley, A. G. "The use and abuse of vegetational terms and concepts," *Ecology* 16:284–307 (1935).

Tobey, R. "American grassland ecology, 1895–1955: The life cycle of a professional research community," in *History of American Ecology*, F. N. Egerton, Ed. (New York: Arno Press, 1977).

Toumey, J. W., and R. Kienholz. "Trenched plots under forest canopies," *Yale Univ. School of Forestry. Bull. No. 30* (1931), 29 pp.

Valiela, I. "Food specificity and community succession: Preliminary ornithological evidence for a general framework," *Gen. Syst.* 16:77–84 (1971).

Van Valen, L., and F. Pitelka. "Commentary: Intellectual censorship in ecology," *Ecology* 55:925–926 (1974).

Vitousek, P. M., and W. A. Reiners. "Ecosystem succession and nutrient retention: A hypothesis," *Bioscience* 25:376–381 (1975).

Vogl, R. J. "Vegetational history of Crex Meadows, a prairie savanna in northwestern Wisconsin," *Am. Midl. Nat.* 72:157–175 (1964).

Walker, D. "Direction and rate in some British post-glacial hydroseres," in *Studies in the Vegetational History of the British Isles*, D. Walker and R. G. West, Eds. (Cambridge: Cambridge University Press, 1970), pp. 117–139.

Webb, L. I., J. G. Tracey and W. T. Williams. "Regeneration and pattern in the subtropical rainforests," *J. Ecol.* 60:675–695 (1972).

Werner, P. "Ecology of plant populations in successional environments," *Syst. Bot.* 1:247–268 (1976).

Whittaker, R. H. "A criticism of the plant association and climatic climax concepts," *Northwest Sci.* 25:17–31 (1951).

Whittaker, R. H. "A consideration of climax theory: The climax as population and pattern," *Ecol. Monog.* 23:41–78 (1953).

Whittaker, R. H. "Recent evolution of ecological concepts in relation to the eastern forests of North America," *Am. J. Bot.* 44:197–206 (1957).

Whittaker, R. H. "Gradient analysis of vegetation," *Biol. Rev.* 41:207–264 (1967).

Whittaker, R. H. "Climax concepts and recognition," in *Handbook of Vegetation Science 8* (The Hague: Dr. W. Junk B. V., 1974), pp. 137–154.

Whittaker, R. H. "Functional aspects of succession in deciduous forests," in *Sukzession Forschung, Ber. Symp. Int. Ver. Vegetationskunde, Rinteln, 1973* (1975), pp. 377–405.

Whittaker, R. H., and S. A. Levin. "The role of mosaic phenomena in natural communities," *Theoret. Pop. Biol.* 12:117–139 (1977).

Whittaker, R. H. "Vegetation and fauna," in *Ber. Int. Symp. Int. Ver. Vegetationskunde, Rinteln, 1976* (Berlin: J. Cramer, 1977), pp. 409–425.

Wiens, J. A. "On competition and variable environments," *Am. Sci.* 65:590–597 (1977).

Wilson, R. E., and E. L. Rice. "Allelopathy as expressed by *Helianthus annus* and its role in old field succession," *Bull. Torrey Bot. Club* 95:432–448 (1968).

Woodmansee, R. G. "Additions and losses of nitrogen in grassland ecosystems," *Bioscience* 28:448–453 (1978).

Worster, D. *Nature's Economy. The Roots of Ecology* (San Francisco: Sierra Club Books, 1977), 404 pp.

Wright, H. E., Jr. "Landscape development, forest fires, and wilderness management," *Science* 186:487–495 (1974).

Zedler, P. H., and F. G. Goff. "Size association analysis of forest successional trends in Wisconsin," *Ecol. Monog.* 43:79–94 (1973).

CHAPTER 2

THE ECOLOGICAL FACTORS THAT PRODUCE PERTURBATION-DEPENDENT ECOSYSTEMS

Richard J. Vogl
Biology Department
California State University
Los Angeles, California 90032

INTRODUCTION

Perturbation is a term used to describe biological or environmental events that produce successional upset, disaster, stress or disturbance to organisms or systems. These events are generally sudden, brief and violent, and are usually considered to be unpredictable and unexpected, even though they are sometimes repetitive. The term perturbation has recently been adopted by ecologists because it has perhaps less anthropomorphic and negative implications than do the terms "disturbance," "catastrophe" or "retrogression" which it has replaced.

Perturbations can be either unnatural or a natural part of a system. Sometimes "stress" is used to designate unnatural perturbations, with stress defined as a perturbation applied to a system to which it is foreign, or a perturbation which is natural to a system but applied at an excessive level [Barrett et al. 1976] or at the wrong time.

The recognition of perturbation-dependent ecosystems focuses on natural events or disturbance factors that are not only inherent to a given system, but are also of such a repetitive nature that organisms and communities have become dependent on them. Man often views perturbations as unnecessary and destructive upsets in nature, but natural catastrophes in perturbation-dependent ecosystems, no matter how violent or harsh, are necessary to the continued existence of the organisms that constitute the system, and are essential to the well-being of that system. In other words, some organisms and systems exist because of certain disturbances or catastrophes and not despite them.

ECOLOGICAL FACTORS

Some of the environmental or biological factors that maintain perturbation-dependent ecosystems will be discussed. Some are widely recognized factors and others are less commonly recognized. A few examples of more subtle factors are included not so much to treat questionable examples, but to demonstrate that these factors can also produce organism control and system maintenance.

No attempt will be made to include all the known factors and perturbation-dependent ecosystems. Such an effort would probably be unsuccessful and only lead to numerous omissions. Rather, the variety of types presented are intended to encourage others to recognize and discover additional factors and examples of perturbation-dependent systems.

Rain and Floods

A variety of systems have evolved to withstand and benefit from perturbations of heavy or excessive rainfall, rapid runoff and floods. These short-lived phenomena have profound influence on certain communities. In Salvador, for example, during a 24-hr period in 1971, over 15 in. (38 cm) of rain fell, and more than 21 in. (53 cm) fell in 48 hr [Cornell and Surowiecki 1972]. Although the lay public usually considers each flood as unexpected and exceptional, meteorologists and climatologists recognize repetitive and even cyclic patterns of unusual rains or floods and often speak of 20-year, 50-year or 100-year rains. Several examples are presented to illustrate some of the ways that rain, floods and sediments act as perturbations.

Salt marsh and estuarine vegetations are composed of halophytes, species adapted to excessive salt and alkaline conditions created by regimes of oceanwater flushing or brackish water accumulations as freshwater flows into and mixes with salt water. The playa salt marshes of the inland arid Southwest and the salt marshes and estuaries of the semiarid coastal Southwest are visited by infrequent rains and are usually devoid of sexual or seed reproduction of the common halophytic species [Barbour 1970; Macdonald 1978]. Only after the occurrence of an exceptionally heavy rain, or a season of heavy rains [such as the winter of 1977–78, when the average season total of 15 in. (38 cm) was doubled in southern California, the second highest annual rainfall on record] does any substantial seed reproduction take place. During these rainy times, colonization and more vigorous vegetative growth occurs. The immediate impact of these exceptional rainfalls is to shock these saline systems with overwhelming injections of freshwater that flush, leach, gut, pond, dilute and flood [Chesapeake Research Consortium 1977]. But the freshwater simultaneously creates the necessary conditions for the rejuvenation, recolonization, spread, genetic diversity, optimum growth and continuation of the inherent plant communities.

Heavy rains or abnormally wet seasons are also responsible for the occurrence of some mesic plant species in more arid regions, and the persistence of mesic species at the drier ends of their distributional ranges. These exceptional rains or wet seasons need occur only once during the lifetime of the species involved for seedling establishment and species renewal, which in the case of trees can be up to 500 years or longer. Big-cone Douglas fir (*Pseudotsuga macrocarpa*) of southern California, for example, apparently has persisted in increasingly arid regions apart from its more northern relative, Douglas fir (*P. menziesii*) by taking advantage of uncommonly wet periods which are often considered destructive or perhaps by a fire that is followed by substantial rains [Bolton and Vogl 1969]. Species like big-cone Douglas fir exist in more xeric regions by their tenacity, longevity and the uncommon occurrence of perturbations that renew life cycles.

In much the same way, some insects and amphibians await the infrequent and exceptional rains that visit local desert regions. The spadefoot toad in the drier portions of the Sonoran Desert around Yuma, Arizona, for example, is known to remain in hibernation for several years until the occurrence of a warm-season rain of sufficient magnitude. These hit-or-miss rains are thunderstorms that must occur in the immediate vicinity of the hibernating toads, be of sufficient duration to soak substrates and also heavy enough to flow, flood and eventually pond in depressions. Again, the seemingly destructive thunderstorm downpours, runoff and resultant floods are the very things that the toads dependently await to reproduce and continue their life cycles.

The winter rains of the Mojave and Sonoran Deserts, although confined to late fall, winter and spring, are inconsistent from year to year. These rains vary in their timing within these seasons and in their annual amounts which are critical factors in determining the kinds, numbers and sizes of the annual plants or wildflowers (desert ephemerals) produced [Went 1948, 1949; Juhren et al. 1956; Tevis 1958a,b; Beatley 1974]. These winter rains are not usually considered to be perturbations because they often lack sudden, violent and destructive characteristics. But perhaps they should be since good wildflower growth years in the desert are exceptional and unpredictable, and the conditions necessary to create these repetitive pulses are the same as those of other recognized perturbations.

In addition to the winter annual plants being dependent upon the proper timing and amounts of rainfall, desert rodent populations are also dependent upon the production of the annual plants for water, forage and seeds, which determines whether or not these mammals will reproduce [Beatley 1969; Chew and Chew 1970; Soholt 1973]. Apparently, certain insect numbers are similarly dependent upon adequate winter rains [Tevis 1958c; Vernon 1975].

Because of the precipitous nature of the terrain and the widespread occurrence of natural pavements that are impervious to water, most desert rains, even when they are not torrential, quickly produce runoff that can gather force and become destructive phenomena. As a result of such repeated catastrophes, the dominant plants that line arroyos or washes formed by floodwaters have become dependent upon them. Smoke tree (*Dalea spinosa*), for example, and perhaps indigo bush (*D. schottii*) and sweet bush (*Bebbia juncea*) do not reproduce unless the seeds that have accumulated in the gravels beneath them are swept up by floodwaters which abrade, wear down, crack and churn the hard-coated seeds in mixtures of sand, gravel, rock and water. As a result of this scarification, the seeds eventually imbibe water, are thoroughly soaked for germination and are distributed to new locations within washes or to newly formed washes, and the seedbed is charged with the necessary water for successful seedling establishment. These same floodwaters often dislodge and destroy the parent plants. In the past 18 years, for example, only three instances of smoke tree reproduction in California deserts have been observed, indicating the irregular occurrence of such floods in the desert.

When the complete desert system is examined, it becomes apparent that the nature of rains, runoff, and floods is not destructive and directionless. For example, sheet erosion, brought about by the actions of rains and winds, removes the finer materials, producing a stone or cobble pavement over much of the surface of desert mountains, bajadas and alluvial fans. This, in turn, assures that rain waters will be repelled, thereby gathering volume and velocity so that by the time the runoff reaches canyons, arroyos and washes it has gained sufficient force to remove the existing plants, which may be suffering from senescence and competition. As the rainwaters gutter and gully, substrate excavation takes place, removing the impervious pavement where present, and even dissecting the caliche layer, a calcium carbonate hardpan formation that sometimes exists below the desert surface. The cutting of the caliche layer, in turn, permits rainwater to penetrate and infiltrate into the alluvial deposits where it can often be reached and used by such deep-rooted phreatophytes as mesquite (*Prosopis* spp.), catsclaw acacia (*Acacia greggii*), ironwood (*Olneya tesota*), palo verde (*Cercidium* spp.) and desert willow (*Chilopsis linearis*). The destructive nature of these floods is a necessity for at least some of the wash species to reproduce and for the survival of the water-pumping species by breaking up hardpans that they could not otherwise penetrate, and recharging their aquifers. Thus runoff, flooding and arroyo-cutting perturbations are apparently part of a self-perpetuating cycle.

Southern California palm oases that exist in desert canyons are also visited by occasional floods that clear out stagnated thickets of arrowweed (*Pulchea sericea*), willows (*Salix* spp.), tamerisk (*Tamarix pentandra*), alder (*Alnus rhombifolia*), reed grass (*Phragmites communis*), three-square (*Scirpus olneyi*) and other rapidly growing species. These infrequent but devastating floods renew water supplies by reducing the number of transpiring plants and recharging the water system. More importantly, the floods clear areas that have become littered with accumulated plant materials for the reestablishment of the California fan palm (*Washingtonia filifera*) which characterizes these oases [Vogl and McHargue 1966]. These events create palm seed-beds of saturated sands in partial shade or full sunlight and also distribute palm seeds and bury them at proper depths in favorable locations.

In August 1976, for example, a flash flood destroyed parts of Pushawalla Canyon Oasis, Riverside County, including the various plants present except for the palms, but by February 1977 over 1000 palm seedlings were present.

Almost all of these seedlings were still surviving during the summer of 1977 and it appeared that there was adequate reproduction to ensure the continuation of this palm oasis for at least another 100 years or longer. Then, as sometimes happens in nature, another, more intensive flash flood occurred in September 1977, destroying all but five of these juvenile palms. At least two more floods followed during the winter of 1977–78. The 1977–78 floods were particularly extensive; almost all plant and substrate obstacles to water flow had been removed by the previous floods.

It is difficult to interpret such seemingly counterproductive phenomena except to point out that even after these repeated floods, in April 1978 there were 653 new palm seedlings in the water-torn and rock-strewn canyon, and more water was present than had been in many decades. In addition, 117 palm seedlings were found up to 3 mi (5 km) downstream of the existing palm oasis. Perhaps this series of floods produced benefits for the species and system, including the possible extension of the oasis. Similar seemingly excessive and counterproductive impacts also occur with other kinds of perturbations which often cloud the chances of accurately assessing their roles.

Perhaps the most dramatic examples of organism and system dependence upon downpour, runoff and flash flood perturbations are found in riparian communities of stream and river systems worldwide. Through the process of hydrarch succession, aided by inputs from the surrounding environments, river systems, including riparian vegetation, progress toward increased nutrients, growth, biomass, eutrophication, siltation, deposition and eventual stagnation. The rates of these inevitable processes are largely controlled by the climate and fertility of the surrounding watersheds.

Generally the only chances for a check or reversal of these successional processes and a removal of the accumulated effects is by catastrophic floods generated by excessive or rapid rainfall, heavy snowfall that melts rapidly or abnormal runoff. Without periodic flash floods, the existing organisms are eventually eliminated and replaced, and the systems often gradually decline in growth and vigor [Ward 1968; Boyd 1970]. These floods often destroy most of the existing plant and animal life, cut old and new banks and channels, scour bottoms of silt buildups and rid the systems of accumulated nutrients.

But at the same time that all this so-called devastation is taking place and real devastation is often occurring to manmade structures, conditions are being created for renewed and vigorous recovery and growth, and the necessary habitats of the declining organisms are rejuvenated. Flash flood perturbations usually act as two-edged swords in riparian systems, simultaneously destroying life while creating improved conditions for the continuation of the very organisms and systems that have been destroyed. It is precisely by these periodic, devastating floods that many streams and rivers remain "eternally youthful" and continue to support the organisms that have evolved in these systems as they receive periodic spurts of vitality through flood perturbations.

Many river systems are maintained from their headwaters to the deltas by perturbations. Rivers slowly but constantly change under patterns of erosion, deposition, accretion, meandering, the formation and eventual death of oxbow lakes and sloughs, the formation of bars and islands and the buildup of deltas. These dynamic processes are countered by a variety of abrupt perturbations that set back and re-energize the systems. They include clearly recognizable perturbations as well as some rather bizarre catastrophic events.

Occasionally, following prolonged droughts or natural draw-downs of water levels, exposed oxbows, cutoff sloughs and peats or sediments accumulated on lake bottoms are ignited by lightning or spontaneous combustion. These fires reverse eutrophication and reestablish more oligotrophic conditions; they are one of the more common ways that the sediments and nutrients in these basins or traps can be removed [Vogl 1977a]. Some of the burned materials leave in the form of smoke, and the remaining ash is subject to removal by wind or by returning water. Reflooding of burned basins often stimulates aquatic plants and animals and results in improved water quality. In some bodies of water, as in the glacial potholes of the Prairie Provinces and States, the seasonal lowering of water levels and abrupt refilling are a stimulating perturbation.

Other examples of perturbation-dependent systems are the deltas that major rivers build, which are periodically lashed by hurricanes resulting in severe wave erosion, saltwater inundation, and freshwater flooding and excavation. The result of this activity is the rejuvenation of the delta system, removal of excessive substrate accumulations and the transfer of nutrients from the terrestrial land mass to the coastal waters. Interestingly, the pulse of nutrients produced by a storm-torn delta or estuary often sets back succession in plankton systems [Lewis 1978]. After the initial shock, however, a surge of new and rapid growth takes place with a response that parallels perturbation-dependent responses in terrestrial systems.

Sometimes floods occur as a result of preceding perturbations such as fires which temporarily remove the vegetation from the surrounding watersheds, resulting in increased runoff. Such fire-flood sequences, for example, are common to most canyons supporting perennial streams in the Southwest. These canyons usually support dense thickets of phreatophytic shrubs and trees and abundant herbaceous and aquatic plants. The long growing seasons, the ever-present water and the temporarily trapped nutrients all promote growth which eventually impairs and stifles additional growth. There is no way for these species to prosper in their own wastes because plant accumulation almost always exceeds decomposition in these canyons.

The deciduous trees of the canyons are particularly important sources of litter. Their leaf accumulations often create continuous trains of fuels that permit higher-elevation lightning fires to penetrate the canyon slopes and bottoms. In addition, tangles of dead and dying vegetation, debris piles created by minor floods and declining water supplies often created by the overabundant plant growth lead inevitably to fires.

Because many semiarid, upland plants do not have the resiliency/recovery abilities of fire-adapted vegetation types, and since the vegetational recovery is delayed because of arid conditions, the precipitous slopes and steep canyons are often subject to rapid, heavy runoff which results in flashfloods, sometimes even with only light to moderate rainfall. The floods that commonly follow these fires tend to be severe, causing massive movement of substrates, even of rocks and boulders, so that the canyon bottoms are scoured, excavated and cut. The finer materials such as the ash, charcoal and organic matter are generally washed down the canyons and deposited on the alluvial fans, floodplains and in the valleys, sinks and playas.

Immediate postfire/-flood assessments reveal multiple damages and negative impacts. The flora and fauna are usually decimated and the canyons gutted. But if the fire-flood episodes are considered from a more extended viewpoint, they can be seen to be part of a cycle that has been repeated innumerable times. This is the way in which these steep-walled canyons were formed and continue to be cut and deepen. The fire-flood sequence is both a temporary end and a beginning by creating favorable conditions for the species involved.

The silts, soils, detained nutrients and accumulated products of photosynthesis that were reduced to ash and scoured out of the canyons appear to represent a real degradation unless each canyon fire-flood sequence is considered in the context of a larger system [Vogl 1977a]. Without a fire there may not have been a severe flood, or at least not one sufficient to clean out the canyon. The fires not only remove the accumulated plant growth that impedes flood waters, but also convert it into bouyant ash and charcoal. This flotsam component is usually transported in the form of an emulsion that resists burial and assures widespread surface deposition. The nonwettable nature of these fire products also contributes to their bouyancy. A flood without a preceding fire will occasionally deliver bulk materials to the valleys but they are usually buried in the soil and/or remain undecomposed because of the arid conditions.

Once these fine materials are spread over intermountain and desert valleys they are vulnerable to wind transport, which could not occur within the protected confines of the deep canyons. Winds commonly rake these open areas, often as "dust devils" and sometimes as violent dust storms or powerful downdrafts preceding thunderstorms. Once these burned plant materials are carried aloft as smoke particulates they can be moved back into mountain watersheds, coming to earth as the winds diminish or as these materials become the nuclei of raindrops and fall. Upon reaching the earth they eventually work their way into and through the canyons again, thereby completing the nutrient cycle. During active thunderstorm periods, which also produce lightning ignitions and downpours, most particulates in smoke plumes or in the air from dust storms do not travel far before they are captured in a thunderstorm and again brought to earth, often on mountain tops.

Water in various forms serves as a key factor in many unrecognized perturbation-dependent systems. Summer thunderstorm showers, as well as the torrential rains, or chubascos, generated by tropical hurricanes may be necessary for the maintenance of the grasslands of the Chihuahuan Desert and Arizona and Sonoran portions of the Sonoran Desert. These grasslands respond dramatically to summer storms, even after overgrazing and when many years have elapsed since the last occurrence of these warm-season rains.

Another possible dependent relationship between capricious rains generated by random thunderstorms and certain organisms may exist on the African veld. There certain birds and mammals seek out storm-affected areas to take advantage of the pulse of vegetation growth and insect emergence triggered by each passing storm. Although sand grouse, pigeons, parrots, thrushes and weavers seem to be dependent on such chance perturbations, the most well known of these "locust" birds is the red-billed quelea. These birds, which form some of the largest flocks in the world, seem to wander about aimlessly until they encounter or "home in" on a recently watered place. The adult birds take advantage of growth stimulated by the rain and nest immediately. The young hatch in 13 days, usually all at once, leave the nest in another 12 days, and these opportunists are off again in search of another rainstorm pulse. Certain mammalian herbivores also trek endlessly across the veld in search of new vegetational growth stimulated by isolated storms. These animals also appear to be perturbation-dependent and, in some cases, may actually be perturbations in themselves.

Wind and Storms

A large percentage of the damage to resources and property that occurs each year because of natural catastrophes is caused by hurricanes (also known as typhoons or cyclones) and tornados. The high winds generated by these and other kinds of storms, including the passage of frontal systems, the presence of high-pressure cells and other uneven or unstable temperature and/or pressure conditions, can be disturbance factors that relate to perturbation-dependent organisms and systems.

The Everglades of southern Florida are an example of a hurricane-dependent system. Periodic hurricane winds and waves literally roll up the coastal mangrove swamps into long windrows of tangled trunks, roots, prop roots and pneumatophores meshed with the fibrous roots of saw grass, turtle grass, manatee grass and the rhizomes of emergent aquatic plants, all of which are draped and plastered with dislodged algae and debris. These "wattle" dams or debris ridges, with the aid of silt and peat, impound the vast expanses of freshwater dumped on the level Everglades by hurricanes, and also hold water produced by normal rains and retard the flow from Lake Okeechobee and the limestone plains in the north [Craighead, personal communication]. The dams also prevent the intrusion of saltwater into the Everglades. At the same time that the dams are formed, conditions are created for the recolonization of the exposed flats by new mangroves, thereby starting the cycle over. The dam network eventually fails, the recolonized mangrove swamps deteriorate, and another hurricane again tears out the mangrove and builds new dams and refloods the flats with freshwater. As a result, the Everglades in general, and the mangrove swamps in particular can be viewed as hurricane-dependent systems.

The total pattern of vegetation in the Everglades is a product of a multitude of disturbances including drastically fluctuating water levels, periodic droughts, frequent fires and occasional frosts [Robertson, personal communication]. Further evidence that the Everglades is a system built upon perturbations, is that despite all of man's stresses and bungling in the Everglades, almost all of the indigenous species of plants and animals are still thriving. In such a system, the less-than-tenacious and nonopportunistic species were probably eliminated long before man arrived.

Everglades hurricanes appear to be destructive to wildlife. Sea turtles are sometimes washed inland where they seem helplessly lost. The Cape Sable sparrow was thought to have been wiped out by a 1935 hurricane but was rediscovered in the 1960s. Wading bird populations also sustain mortality from hurricanes. But despite these losses, most experienced field biologists admit that the improved habitats and increased species numbers produced because of hurricane perturbations outweigh any of the immediate losses, although few long-term studies have been conducted to document the necessary and beneficial aspects of hurricanes to the wildlife of this region.

The barrier islands of the Outer Banks and the sand spits of other Atlantic Coast locations are also examples of hurricane-dependent ecosystems. In these areas periodic hurricanes produce high seas that overwash the sand communities, thereby removing the established vegetation in certain sections and excavating the accumulated sands. At the same time that the displacement of plants and substrates is taking place, the conditions for rapid recovery are being created which often include the deposition of plant fragments which help to ensure quick recolonization.

The characteristically unstable nature of sand, whether it is above or below water, contributes to its rapid and easy removal, but also facilitates its ready return and accumulation. The removal of the stabilizing vegetation allows the wind to again interact with the exposed sands and the colonizing plant growth to rebuild the dunes. In some cases the barrier islands or sand strands are breached by the storm waters which results in the creation of new inlets to the salt marshes that often exist behind the islands or that flank the protected waters of the sounds. In some instances, these new inlets permit an increased salt water exchange (tidal prism) which apparently causes renewal and enrichment of the salt marshes and the sound fisheries. This, in turn, usually results in increased bird utilization and productivity [P. J. Godfrey, personal communication].

The returning dune-building plants in the overwash or scour zones almost always grow more vigorously than those plants that were removed. Many of the established plants slow up growth or become senescent and often stagnate, with little or no chance of renewal without wholesale disturbance. The grass, herb and scrub vegetation of these sand islands, strips and spits, as well as perhaps the maritime forests that eventually develop on their inland sides, are therefore renewed by these periodic overwash disturbances. In addition, stability, quasi-equilibrium and successional changes cannot be understood in these communities without first recognizing the cyclic dynamics of these build and tear systems.

Colonial sea birds such as terns are also dependent upon the newly formed areas of overwash fans, sand spits and other sand accretion areas for successful nesting sites. Densely vegetated areas or large dunes formed by sand-entrapping plants are generally unsuitable for birds whose nesting and resting (loafing) sites require relatively clear horizons at bird level. Apparently, recent efforts by man to stabilize and fortify Atlantic coastal dune systems have contributed to the decline of some of these colonial bird populations. The hurricane creation of new inlets and the enrichment of sound waters also contributes to successful nesting by increasing food supplies in the immediate vicinity of sparsely vegetated nesting sites.

The number of hurricane-killed trees and large shrubs in the maritime forest, usually produced as a result of saltwater poisoning, also tends to regulate the numbers of colonial nesting marsh birds such as herons that depend upon these dead woody species for rookery sites. Many birds of the sandy seashore and marshlands, as well as certain littoral marine organisms, have been reduced in numbers and/or have had to seek alternative and usually less favorable sites for their continued survival as man's actions have mitigated the impacts of natural disturbances and have impaired the system dynamics by stabilization.

Coastal strand ecosystems are relatively simple systems supporting few species over a generally homogeneous geomorphology. As a result, hurricane perturbations help to increase the total diversity and to maintain it by interrupting the normal morphological uniformity in space and time and by creating certain features and habitats. The numerous overwash fans of the low dune barrier islands around Cape Lookout, North Carolina, for example, create favorable conditions for a number of additional plants and animals. Some barrier islands in the Gulf of Mexico contain beach ridges that are distinct from sand dunes in that these ridges are derived from hurricanes and are composed of marl and shell fragments in addition to sand. These beach ridges tend to be more persistent than the wind-built sand dunes and are dominated by a distinct herbaceous vegetation that is not readily invaded by shrubs and trees [R. K. Godfrey, personal communication].

Recent data obtained from four barrier islands in the Gulf of Mexico relate hurricane incidence to the establishment of slash pine (*Pinus elliotii*) stands [Stoneburner 1978]. On these strands it appears that hurricane-generated overwash has replaced fire as a perturbation by creating the proper pioneer conditions for slash pine reestablishment at the same time that mature and overmature pine stands are decimated and unfavorable plant competition is reduced.

Pond pine (*Pinus serotina*) and sand pine (*P. clausa*) may also reproduce after coastal stands of these southeastern U.S. species are lashed by hurricanes and shocked and poisoned by saltwater. Interestingly, coastal strand locations around North America's coastlines support stands of closed-cone pines such as beach pine (*Pinus contorta contorta*), bishop pine (*P. muricata*), Monterey pine (*P. radiata*), pygmy pine (*P. contorta bolanderi*), Torrey pine (*P. torreyana*) [Vogl et al. 1977], pitch pine (*P. rigida*), pond pine, and sand pine, as well as several closed-cone cypress (*Cupressus*) species. These serotinous-coned species are usually dependent upon fire for the release of seeds from their cones and for successful reproduction. In some coastal locations these closed-coned species are joined or replaced by other fire-dependent pines such as longleaf pine (*P. palustris*) and loblolly pine (*P. taeda*). But in at least some of these sand communities, fires appear to be very infrequent, perhaps absent, and reproduction seems to take place without their occurrence. This predominance of fire-dependent conifers along North America's coastlines cannot be considered to be just coincidence, but must somehow relate to the occurrence of coastal perturbations that create effects, selectivity and dependence that are a close ecological equivalent to fire.

Strong desert winds are unpredictable, although in some regions they are more common in certain seasons and are often directional for given periods. Living desert sand dune communities are dynamic sand, water, organic matter and seed traps that are operated by and dependent upon vagrant desert sand storms. Sand must be moved by strong enough winds to build dunes, a process aided by certain plants. The accumulated sand enables the dunes to retain rainwater and maintain zones of capillary and vadose waters. The wind-sorting of sand, clay and silt particles in the dunes facilitates the formation of incipient, impervious hardpans when it rains, enhancing the dune's water storage capacity.

The entrapped water aids the establishment and growth of plants, the fermentation of organic matter and the growth of fungi, and provides subterranean refuge from the hostile desert environment for certain insects, herptiles, snakes and small mammals. These organisms get most of their food directly or indirectly from the accumulated seeds that the winds have scavenged from immediate and distant areas and deposited in the dunes. All of these processes and activities are far more complicated than the oversimplifications presented here to demonstrate the dependence of desert dune systems on wind. Without these periodic wind and sand storms, the dune systems would soon run down, and the plant and animal life they support would steadily diminish, and the termination of the winds would lead to static and impoverished systems. Coastal sand dunes, in contrast, are usually impoverished because winds sweeping over the oceans cannot gather seeds, and inland dunes without winds would similarly support few or no animals.

Most dune systems are not only maintained by wind disturbance, but were originally created by perturbations associated with glaciation. These included excessive rain and snowfall, massive erosion of desert mountains and the formation of Pleistocene lakes. The outwashed materials that finally reached the valley and lake bottoms were subject to wind- and storm-generated wave action that eventually piled the sands in the vicinity of the present-day dunes.

A number of sun-loving herbaceous plants and shrubs are dependent upon wind along with secondary agents such as insects, fungi and lightning to fell individual or groups of trees to create light gaps in forests. These gaps are an inherent part of tropical to boreal forests and help maintain diversity of age and species in the otherwise dense, continuous and homogeneous forests. Without periodic strong winds to create these small and scattered openings, certain opportunistic species would be without suitable habitats.

Some birds and mammals, in turn, are dependent upon the productive pioneer plants that rapidly crowd these openings, thereby contributing to faunal diversity [Thompson and Willson 1978]. When these small clearings and temporary wildlife centers in the forest are considered as subsystems of the larger forest tracts, their dependence on wind is apparent. Even the characteristic tip-up mounds and depressions created by upheaved roots when overmature trees are windthrown create habitats for certain species [Vogl 1973].

When the whole forest ecosystem is considered, gaps created by winds take on new dimensions. The seemingly randomly created openings are often not random at all, but are a systematic way in which forest diversity is maintained by the creation of a mosaic pattern of different-aged groups. The winds act like a predator by taking out the larger and usually overmature and senescent individuals and returning them to the forest floor where they can be recycled. These winds, then, which are looked upon with disdain by some foresters and lumbermen, are essential factors in forest turnover and maintenance.

Wind, like most other factors, has not been completely explored as a controlling factor in relationship to dependent systems [Vogl 1973]. Tornados, for example, occur repeatedly in definite "alleys" or belts that spread across continents. Plant and animal communities are affected by these tornados and even human lives and activities are influenced, not only by the damage produced by the tornados, but also by the more subtle changes in air ionization and the stimuli of abrupt changes in temperatures, humidities and pressures that occur during the tornado season. Surely, there must be more than just casual relationships between the organisms and systems and the tornados that have plagued people for thousands of years.

A recently described wind phenomenon called a downburst may play a subtle but controlling role in the northern Lake States [Taylor 1978]. A downburst is similar to a tornado in its strength and effects, but without the twisting action. This type of storm involves outbursting winds from downdrafts that are avalanches of air that strike the ground at high velocities. A July 4, 1977 downburst in Wisconsin consisted of 25 separate blasts of air with winds up to 157 mph (253 km/hr). These occurred in a swath 17 mi (27 km) wide and 166 mi (267 km) long in which almost all the second-growth forest, as well as part of a virgin forest, were leveled. Although this is the first such storm to be studied by meterologists, such big "blows" have occurred twice before in this century and in nearly the same area. The 1949 blowdown was followed by at least 10 years during which time ruffed grouse and white-tailed deer numbers erupted in a response to the opening up of the forest and the invasion of fruit and seed-bearing herbs and shrubs. Perhaps such animal species are dependent upon the occurrence of these timber blowdowns for optimum conditions, and other species are actually dependent upon what seem to be totally destructive and random perturbations.

Fire

Fire is recognized as one of the most common perturbations to disturb and traumatize various ecosystems. Almost every ecology textbook presents fire as an example of a natural catastrophe or a retrogressive agent that upsets and sets back the vegetational development. More recent texts include examples of natural systems that have become adjusted to fire and discuss organisms that have developed recovery mechanisms and fire adaptations proportional to fire frequency and intensity. Some systems are fire-dependent, possessing organisms that require fire for their survival, well-being and continuance, with fire an essential part of the environment.

The dominant plant species in fire-dependent systems are not only adapted to fire, but also possess fire-dependent structures, mechanisms and functions. These include such things as postfire seedbed requirements, fruiting bodies that release seeds upon exposure to fire and seeds that remain dormant or stored until treated by fire. In most of these plant species, sufficient reproduction does not take place without fire, making them obligate fire-types. In a few species such as wire grass (*Aristida stricta*) of the Southeast, and knobcone pine (*Pinus attenuata*) of southern California, sexual reproduction is controlled exclusively by fire treatment [Vogl 1972, 1973].

The majority of animals in these systems, in turn, depend on postfire vegetation/habitats, and function in ways to survive and take advantage of the periodic or frequent fires. The Kirtland Warbler, for example, depends on young jack pine (*Pinus banksiana*) forests that are produced and maintained by fire. The greatest arrays of higher animal species and the highest densities almost everywhere in the world are associated with fire-affected ecosystems, with the highest animal numbers usually occurring in the early postfire stages. The great herds of herbivores, with their predators and scavengers, of every continent are associated with fire-dependent grasslands and savannas [Vogl 1974].

Fires often serve as the driving forces of the life cycles of the plant species present in fire-dependent systems. They maintain the vegetational development at given stages. Sometimes fires may cause and also be the result of the existence of certain plant species. Fire often serves as the important decomposition or nutrient recycling agent. Fire-type communities usually produce abundant fuels that accumulate faster than they decompose because of decay resistance and/or climatic conditions. These plant accumulations are generally very flammable or become flammable before burning. The physical and chemical compositions of the plants can also enhance combustion.

Some fire-dependent species have adaptations to resist fire mortality such as ground-level or below-ground meristems like those found in many chaparral and grassland species. The reproductive bodies of some species are relatively fireproof. Others have thick bark or protected meristems, common adaptations of dicot and monocot trees of savannas around the world. Species that are damaged by fire possess the ability to survive defoliation, produce epicormic stem sprouts or produce root or root-crown resprouts or suckers.

Fire-dependent species, whether they are killed by fire and then reproduce by seed, or survive fire in some manner, are characterized by rapid growth, precocity, high reproductive rates, short lifespans and even-aged groups of the same species. Regrowth and reappearance is often phenomenal after initiation by fire and associated environmental changes. The postfire response is actually a rejuvenation as well as a recovery, in that the fires often initiate new life, stimulate growth and promote vigor.

Fire-dependent systems occur in regions of natural ignition. Climatic conditions, landforms and other environmental factors are also conducive to frequent fires. Examples of fire-dependent ecosystems are found in a variety of climates and vegetation formations [Vogl 1977b].

Fire-dependent plant communities can be subdivided into those that are maintained by fires and those that are initiated by fires. Fire maintenance occurs when frequent (every 1–10 years), low-intensity fires prevent excessive fuel buildups which might result in severe fires that would destroy the existing vegetation. It also controls plant competition, invasion and succession which would result in vegetational replacement. Fire-intolerant species are selectively checked and eliminated. Fire is the principal decomposing agent since the fuel composition and environmental conditions in these communities deter decomposition by bacteria, fungi and invertebrates.

Repeated low-intensity fires, which are usually surface fires, selectively thin, prune or remove the standing crop and control the invasion of competing species. Many fire-maintained plant species are prolific seeders, and establishment and recruitment are regulated by recurring fires. The fires also tend to eliminate the slow-growing individuals, those growing on poor sites and the genetically inferior ones.

Many regions capable of supporting forest are converted by frequent fires to savannas, that is, grasslands with an open overstory of widely spaced trees, or parklands of dense groves of forest and/or brush surrounded and separated by open grasslands. The fires in these savannas control woody plant structure and development, and maintain the grasslands by removing plant accumulations which physically impair growth by depriving the grassland plants of space and light. Savannas in various parts of the world are maintained by repeated fires, and grasslands, ranging from xeric to wet types, are also often dependent upon fire maintenance [Vogl 1974]. Certain shrub communities such as California chaparral are renewed by periodic fires which maintain vigorous growth, and species and age diversity through the creation of mosaic patterns.

Fire-initiated systems are dependent upon infrequent catastrophic fires that simultaneously terminate and initiate long-lived species. During the period when the system is free from severe fires, which may be for several hundred years, mesic species become more and more important as the less mesic, fire-initiated species are reduced by competition, lightning, wind and insects. The surviving individuals, however, usually create conditions that lead to an eventual fire which destroys them along with the associated species, but concurrently creates conditions necessary for the reestablishment and continuance of the species and system. Little or no successful reproduction of these fire-initiated species takes place during the fire-free period; at best, reproduction does not keep pace with mortality. Mortality of the established individuals is high in the early postfire years but quickly declines to very low levels. The fire-initiated plant species are usually shade intolerant, grow rapidly until reaching maturity and are long-lived and tenacious. Once established, these pioneers modify the microenvironment and more mesic nonfire species begin to appear. Fire-initiated systems are common in temperate and boreal regions [Vogl 1977b].

The relationship and dependency on fire of these fire-initiated systems is often subtle and unrecognized. The role of fire is obscured by the long life-spans of the species involved, the infrequent occurrence of the fires (once every 100 to 1000 years), the token reproduction that occurs without fire, the association with nonfire species, and the former considerations of these fires as being only negative and destructive forces. The dependencies of organisms on other types of perturbations are also undoubtedly unrecognized because of similar circumstances.

Snow and Frost

Snow and frost can act as controlling factors in various communities in high-altitude and high-latitude environments. Survival and reproduction of some arctic and alpine small mammals, for example, depend on snowdrift accumulations. Certain woody plants in these same communities are undamaged by frost and winter "burn" or desiccation when protected by snow accumulations. In some areas the presence or absence of woody plants at the edge of tundra is determined by the presence or absence of winter snowdrifts. Willows, ericaceous shrubs or krumholtz conifer tree species are often snow-dependent at treeline, not only on the protective cover of snow, but also on the critical moisture that it provides as it melts during the growing season.

Snow often acts as a perturbation factor in boreal forest regions. A substantial snow at the onset of the growing season will often insulate the soils against frost, thereby leaving the shallow, plate-root systems in soft and loose substrates. If heavy, wet snow adheres to the branches of these conifers and strong winds accompany or follow the wet snow, a widespread timber blow-down can take place because of the unfrozen soils and the top-heavy trees. The snow felling of mature spruce/fir stands, like fire and/or insects, serves to terminate the existing growth, but also sets up the necessary conditions for the renewal of the same species. Sometimes several of these perturbations act together in sequence.

The community of organisms occupying the numerous avalanche chutes of the upper elevations of northern mountains also depend on periodic snow slides. In these areas of natural conflict between snow and forest, communities of pliable shrubs, small trees and herbaceous plants often dominate. Although these species usually occur elsewhere, without the destructive force of periodic snow avalanches these communities would be replaced successionally by coniferous forest.

Erosion

Some locations and regions are particularly susceptible to rapid and continual erosion because of certain climatic features and substrate types or structures. In places where unstable sites have existed for substantial periods of time, organisms often also exist that have become dependent upon them. Erosion-dependent organisms most commonly occur in such places as deserts, where the climate has not allowed the development of a stabilizing mantle of vegetation; in areas of precipitous terrain created and maintained by active mountain building or areas subject to intensive erosion because of particularly friable substrates, peculiar locations or because of extreme environmental conditions. In a few instances the organisms dependent upon these unstable sites are unique, but most often they are sun-loving pioneers that occur elsewhere in very low numbers, but thrive and proliferate on active erosion sites.

A few examples might help to illustrate the dependence of certain organisms and communities on erosional processes. In arctic tundra regions, steep terraces or escarpments formed during glacial times which flank major river floodplains continue to remain raw and active as freeze-thaw actions cause these slopes to slump and cut back by the process of cryoplanation. These sites, as well as other places of permafrost thaw and slippage, are usually composed of seepy glacial till and flour and support only certain annual and short-lived weedy plant species that cannot generally grow on the normal peaty soils and compete with tundra shrubs, herbs and grasses.

Other examples of erosion-dependent organisms are the species of *Gunnera*, a gigantic herb of tropical montane cloud forests, occurring mainly in the southern hemisphere. These species dominate nonwoody plant communities that occupy landslide sites [Palkovic, personal communication]. These herbs cannot survive in the dense cloud forests and have persisted only by growing on the open, unstable and seepy slide areas created as heavy rains produce water-logged clay soils on steep mountain slopes that eventually succumb to gravity. Recent encroachments of man into parts of these tropical forests have caused an expansion of the populations of *Gunnera* species by the creation of road cuts, logged sites and forest clearings. These disturbed areas are readily invaded by these pioneer species, thereby reducing the exclusive dependence of *Gunnera* species on landslide sites.

A more subtle example of erosion dependence is the occurrence of knobcone pine (*Pinus attenuata*) in the Santa Ana Mountains, California [Vogl 1973]. In these mountains, knobcone pine is confined to a serpentinite outcrop near the summit of a mountain peak. The controlling influences of the serpentine diminish, however, as the total plant growth interacts with the parent material to form and hold soils and produce litter. But these processes are checked by continuous erosion, which is periodically accentuated by fire augmented by wind-throwing the dead trees and uprooting of fresh serpentinite. The serpentinite itself is extremely friable and subject to rapid weathering. In addition, these isolated, hydrothermally altered serpentinite bodies are usually associated with active faults so that they are uplifting or expanding at a more rapid rate than the surrounding areas, thereby ensuring steep slopes and a continuous supply of fresh serpentinite with erosion. The continual erosion tends to counter the modifying effects of the vegetation and retards plant invasion from the surrounding areas. Knobcone pine is believed to have survived on this isolated geological formation as a result of balanced interactions between geological activity and erosional instability. This perpetuates the pioneer conditions necessary for the knobcone pine continuance and controls the competitive plant growth that would otherwise eliminate the pines. In addition to being dependent on a specialized substrate, the knobcone pine is also dependent upon fire for reproduction [Vogl 1973; Vogl et al. 1977].

Mountain Building, Volcanism and Glaciation

Although the great earth-shaping events that formed the earth's surface occurred most actively in past eras, these large-scale perturbations have left lasting effects that are still influential and determinant [Cornell and Surowiecki 1972; Ward 1978]. In addition, mountain building, volcanism, glaciation, erosion and water action forces still occur actively in some areas, or will again resurge since these processes are repetitious events on long-term cycles. These various earth-shaping processes determine the geomorphology, which in turn reacts with the climate to dictate the kinds of soil, water, plants and animals that a given area will support. Some of these long-term natural events, although nearly unnoticeable, may have a greater effect on the future of mankind in the long run than the obvious perturbations of contemporary systems [Fuchs 1977].

In northern latitudes, for example, old glacial lakebeds still contain sandy, infertile plains that support unique pine and oak barrens that do not occur elsewhere [Vogl 1964, 1970a]. In the far north, such systems as boulder-train streams nourished by mountain glaciers, and glacial-fed rivers with milky waters laden with glacial "flour" support specific communities of organisms. Systems like alpine tundra, even those areas completely released from the icy grip of glaciers, are still profoundly influenced by the moraines, cirques and fellfields formed by those glaciers.

Large desert sand dune systems, formed during wetter periods in inter-glacial times by the powerful erosional forces of water and wind, often occur on the former lakeshores of the now dry lakes and support unique fauna and flora [Stebbins 1974]. Even the fairy shrimp (*Branchinectra* spp.) and other primitive crustaceans of the desert dry lakes or playas of the Southwest become active only when current climatic pulses are reminiscent of former conditions when these lakes were first formed and active. In both examples, the important factors affecting these desert systems were created in past times, but are still effectively controlling them.

Certain organisms and systems are uniquely adapted to volcanic cinder cones, lava flows and volcanic ash beds and are dependent upon these geomorphological formations even though they were formed eons ago. Such things as kipukas, islands of mature and unusual vegetation surrounded by lava flows, are dependent upon repeated volcanic disturbance for their production and maintenance. Species such as tree ferns (*Sadleria* spp.), certain Pteridophytes and other lower plants may be dependent upon active steam vents, fumaroles, lava tubes, volcanic gas disturbance, ash stimulation and volcanic heat and fire [Vogl 1969b]. In addition, volcanic eruptions may sometimes cause global perturbations. Studies of the recent eruption of Mount Agung on the island of Bali, for example, revealed that gas, smoke and tephra girdled the globe and caused significant climatic changes [Hansen et al. 1978].

In some tropical and subtropical areas, the continual leaching of volcanic substrates by rain for hundreds of years has left abnormal concentrations of iron in the soils. As a result, many of the existing plants in such areas must be able to resist or cope with the iron toxicity. The iron concentrations are a controlling condition that was initiated during the last volcanic eruption and brought to completion by heavy rainfall that robbed the lava substrates of almost all minerals but the water-resistant iron [Fosberg, Lamoureux, and Swindale, personal communications].

Another natural force that often has organism and system dependencies is water. Because of its omnipresent nature, it is seldom considered as a catastrophic agent unless floods or torrential rains are considered individually. And yet, many systems are developed and maintained by water relentlessly, periodically or cyclically working and wearing century after century. Water produced by precipitation and powered by gravity, winds and tides untiringly creates canyons, including dalls and dells; cuts river channels; builds eskers and moraines; forms alluvial fans; breaks down mountains; transports and sorts substrates; and shapes seaside bluffs, various kinds of beaches, sand dunes, floodplains and terraces.

Existing organisms and systems can therefore be dependent upon earth-shaping events that occurred in the distant past. When water, erosion, glaciation and volcanism are considered on a geological time scale, we tend not to consider them as perturbations or catastrophes. But when events regarding these forces occur today, we seem often to forget that they are part of long-term dynamic processes and we usually view them as short-lived phenomena and classify them as clear-cut perturbations.

Animal-Caused Perturbations

Certain plants and vegetation types are maintained and promoted by the activities of wild and domestic herbivores. Browsing mammals in such places as wintering grounds, yards and ancestral home areas are known to control and stimulate palatable woody plants. Without continual or repeated browsing these woody plants would soon be out of reach and no longer available for such mammals as elk, deer and goats. The removal of terminal buds by browsing often stimulates the growth of the latent and lateral buds and/or root and rootcrown sprouts. In other words, moderate browsing often creates and perpetuates browse, and keeps the browse plants within reach of the animals. Elk and moose in Prince Albert Park, Saskatchewan, for example, have maintained distinct openings in the otherwise dense boreal forest by regular browsing. Many stands of old-growth curlleaf mahogany (*Cerocarpus ledifolius*) in the West have provided browse for wintering deer and elk since the time of white settlement, even though seed reproduction has not been known to occur during that time [Young et al. 1978].

In some cases, herbivore browsing maintains shrubs and trees in the early stages of development, often preventing them from reaching the point where they mature or become physiologically decadent. The active plant growth, in turn, supports the herbivores that are necessary for its maintenance.

Large grazing mammals, both wild and domestic, also maintain grasslands under certain conditions. Grazing animals remove the excess plant growth that would otherwise accumulate and physically and chemically impair further growth [Vogl 1974]. Light animal trampling often facilitates successful establishment of grass and forb seedlings, and seeds of some species are spread and perpetuated by grazing animals. Many grassland openings in forests in the Intermountain West, for example, occur on steep southwest exposures with shallow soils, but are usually maintained only by heavy animal utilization [Ffolliott et al. 1977; Singer 1979]. Some African veld big-game species are known to control the plant composition and structure of certain grassland/savanna areas [Wilson and Hirst 1977; McNaughton 1979]. The encroachment of woody species onto grasslands is not only checked by repeated fires and large animals [Vogl 1974; Naveh 1975], but also by a variety of small herbivores, including rodents and other small mammals.

The establishment, spread and maintenance of such recognized vegetation types as cactus scrub, sage scrub and thorn shrub communities are often the result of close grazing by wild and/or domestic herbivores. Even birds such as wild geese and certain puddle ducks can maintain grasslands and sedge meadows by their grazing actions. In some cases, their intense utilization will maintain these areas in seedling stages by the continual uprooting of the grasses and sedges.

Certain pelagic and other seabirds that nest in colonies on islands, headlands and rocky bluffs, and some marshland birds that establish rookeries in trees, often appear to control and maintain their nesting sites. A basic requirement of these birds is that the nesting sites remain relatively free of plant growth so that the birds have open vistas and can view other nesting birds. This requirement is apparently necessary for the colony to function properly. In addition, open nesting sites greatly reduce losses from predation and also ensure unobstructed and easy ingress and egress for birds that are usually awkward and clumsy in their landings and take-offs. Some colonial birds like the endangered least tern (*Sterna antillarum*) of southern California, for example, sometimes obtain the necessary openness by nesting on playas or salt flats which are plant free because of the high salt concentrations that occur there. But some colonial seabirds and marsh birds actually create and/or maintain open sites by producing large quantities of guano that persist and accumulate. These concentrated nitrogenous wastes kill the existing plants and check plant reinvasion and succession. Trees that are killed in this manner are quickly defoliated, but the woody portions persist for long periods, thereby making ideal open nesting platforms. If high concentrations of animal wastes can be considered a disturbance factor, then functional nesting sites are maintained by these animal-caused perturbations.

Beavers in many North American forest locations with lakes, rivers and streams are classical examples of retrogressive agents whose destruction/disturbance checks and reverses plant succession and determines the stage of vegetational development. Aspen, cottonwood (*Populus* spp.) and birch (*Betula* spp.) stands are renewed and maintained by beaver cutting or girdling which creates new sucker sprouts and thereby circumvents senescence. Without disturbance the tree stands are replaced successionally by other species. Trees that are eliminated by beaver flooding and harvesting often reinvade these former sites after the beavers abandon their dams again exposing moist, bare soils. The flooding of shrub and herb communities by beaver damming and the subsequent draining of these areas often also creates additional areas for deciduous tree establishment. In many areas, streamside and lakeshore stands of palatable deciduous forest trees are controlled and dependent upon the presence of beavers. The activities of the beavers, then, are a vital link in the life cycles of these tree species.

In assessing the dependence of certain plant or animal communities on the activities of animals, the definition of a perturbation or disturbance comes under scrutiny. Predators, for example, often control prey populations, and the prey are dependent upon the predators for their healthy maintenance. But should predation, such as a pack of wolves culling out the weak, unfit and old moose be considered a disturbance? Perhaps it should, because if the same animals were killed by a severe storm, for example, the resulting mortality would be considered a perturbation. If biological/environmental occurrences must be particularly destructive to be considered as disturbances, then the degree of the destruction becomes a key issue.

Sometimes short-lived phenomena and other catastrophic occurrences are considered perturbations because they appear to be abnormal events. But if the time frame in which these "unusual" events occur is expanded, the events are often seen to be repetitive or cyclic. Then, the so-called catastrophes often turn out to be necessary occurrences and are as normal as are such things as predation, pollination and seedling establishment.

Insect and disease outbreaks are often considered as natural catastrophes even if they are cyclic. These epidemics are recognized as relating to certain environmental and host species conditions.

The lodgepole pine (*Pinus contorta murrayana*) of the Sierra Nevada, for example, has been subject to periodic outbreaks of the lodgepole needle miner (*Coleotechnites milleri*) which seems to follow a 2-year cycle [Johnston 1970]. In a 10-year buildup period, for example, mature trees may be almost completely defoliated. Three major outbreaks during this century among mature and overmature lodgepole pines have killed thousands of trees, but each time young lodgepole pines replaced the dead ones. Although the details of the relationships between the needle miners and the lodgepole pine are not fully known, it appears as if the lodgepole pine communities of the Sierra Nevada may be dependent upon these periodic insect epidemics for their renewal and continuance. This type of perturbation dependence parallels the dependence of the Rocky Mountain subspecies of lodgepole pine (*Pinus contorta latifolia*) on fire. The Rocky Mountain lodgepole pine cones must be opened by fire for successful reproduction. Sierra Nevada lodgepole pine cones, however, open upon maturity, but both varieties similarly require catastrophic agents that destroy old individuals for the renewal and continuance of the species and the system.

The young Sierra lodgepole pines support ever increasing populations of leaf miners along with parasitic wasps and bird populations, particularly chickadees that feed heavily on leaf miners as well as on lodgepole pine seeds [Johnston 1970]. Similar system and organism dependencies on insect outbreaks undoubtedly exist, but are not now known because only the destructive or negative aspects of these phenomena have been considered.

Another example of an animal that may control or help to maintain a vegetation type is the desert termite (*Gnathamiteres tubiformans*) of Southwest semiarid grasslands, desert grasslands and deserts. Although these insects are often abundant, little is known about their life cycles and food habits [Allen and Foster 1977]. Desert termites utilize both living and dead plant materials. In some grassland areas the termites are present almost every year with occasional peak years, while in the more arid desert areas, outbreaks of these insects are infrequent. In these desert regions, dead woody plant parts in contact with the ground, grasses, forbs and annuals are often totally consumed during outbreaks. In these dry environments the usual decay agents such as bacteria and fungi are normally inoperative, and the vegetation is often too sparse to carry fires. It may be that periodic termite outbreaks are essential in the turnover of nutrients locked in the standing dead crop and accumulated litter. These nutrients may be critical to the continued well-being of these systems because of the total low amounts of nutrients present in these impoverished and infertile systems.

In relationship to insect dependencies, Peterman [1978] considered outbreaks of mountain pine beetles (*Dendroctonus ponderosae*) as a natural harvesting and thinning agent of *Pinus contorta latifolia*. The pine beetles may not only create fuels for the next fire, but could also decrease the number of trees and accumulated cones that will be present at the time of the fire. This might mean that fewer seeds will be released by a fire, thereby preventing the establishment of dense thickets of lodgepole pine that quickly stagnate. Peterman [1978] concluded that outbreaks of mountain pine beetles actually regulate and benefit lodgepole pine rather than act as detrimental pests. It may be that under certain conditions lodgepole pine systems are dependent upon periodic insect as well as fire disturbances.

Sometimes insect epidemics relate to systems in a reverse way. Monocultural croplands or degraded second-growth forests [Lang et al. 1978] as a result of man-caused disturbances, for example, invite buildups of insects and diseases. In these instances the role of the insect outbreaks is reversed because the insect buildups are caused by man's disturbances and the establishment of monocultures with the vegetation controlling the insects instead of the insects controlling the plants.

The various ideas and examples in this section were presented, not so much as absolute cases of dependence on animal-caused perturbations, but to focus attention on animals as perturbation agents, and to present the viewpoint that some insect outbreaks are necessary events in the system.

Man-Caused Perturbations

Most physical disturbances introduced by man are too sudden and violent for adaptation or adjustment [Odum 1969]. But some of the natural communities altered by man, or systems created by him, depend upon man's repeated intervention for their maintenance. Most of these systems need no explanation because of the obvious relationships between the man-caused disturbances and the maintenance of the various systems. The relationships are usually simple cause and effect types, such as lawns which are maintained only by continued care and mowing, with the most vigorous growth promoted by repeated and consistent mowing disturbance.

Important systems that require man's repeated disturbance and occupy large areas are the various agricultural operations, ranging from primitive slash and burn agriculture and small gardens to modern farming, including giant agribusiness complexes [Anderson 1967]. Within agricultural systems, the cultivars are not only dependent upon plowing, pruning, cultivating and cutting disturbances, but so also are some of the weed plants, plant diseases, insects, birds and mammals, including some widespread pest species. Rib grass (*Plantago* spp.), for example, appeared in Denmark along with Neolithic farmers and has been a widespread agricultural weed ever since [Anderson 1967]. Agricultural perturbations are not restricted to terrestrial sites, but occupy aquatic sites converted to such things as fish ponds, rice paddies and taro patches. Many of these systems only sustain their productivity by such disruptive disturbances as periodic drawdown and reflooding, along with cultivation and replanting.

Beyond such things as orchards, tree plantations and permanent pastures, most agricultural systems are maintained as annual build and tear operations. In addition, the monocultures that man usually creates by his repeated and deliberate perturbations often lend themselves to, or invite, further disturbance by natural catastrophes. Modern farming practices, for example, often invite hordes of pest or "locust" birds. These catastrophic crop destroyers are perhaps indicators of agriculture's overreachment since such reductions in diversity usually result in instability. These latter perturbations are often totally destructive from man's standpoint, but are, in actuality, responses of nature as a result of altering the original systems.

In various parts of the world, particularly in the tropics and subtropics, anthropogenic grasslands have been created as a result of man's activity in regions that normally support brush or forest [Baker 1971]. The brush or forest is removed and the grassland maintained only by repeated cutting and burning as well as by actions of domesticated and feral livestock [Vogl 1974]. Grazing, trampling and compaction of both natural and manmade (anthropogenic) grasslands create conditions that encourage and promote grasslands and are characterized by certain annual weeds, nonpalatable plants, atypical spreads of certain shrubs, high rodent populations and unstable conditions. Continued grazing usually effectively promotes and maintains these altered and degraded communities.

Many tropical areas that have been severely disturbed by man are now dominated by bamboo (*Phyllostachys* spp.) or bracken fern (*Pteridium aquilinum*). Bamboo, interestingly, terminates life after flowering, thereby almost ensuring disturbance renewal by fire.

Feral dogs and cats, cockroaches, house flies and Norway rats are obvious examples of organisms dependent upon man's activities. In many respects, man himself is a boom and bust type, dependent upon his repeated disturbances for survival [Odum 1969; Odum et al. 1979]. Less obvious are those organisms that respond to man's dumps and trash heaps, the unusual assemblages of common plant species, or those rare species that flourish in such places as early-man kitchen middens, mine sites and ghost towns [Baker 1970]. The only sites on north Florida Gulf Coast barrier islands that support magnolia-beech forest, for example, are located on ancient oyster shell middens, illustrating the lasting influence of man [Clewell, personal communication]. The fulmars that traditionally nested in the Arctic spread south during the past century into British and Scandinavian waters following the man-produced foods of whaling ships and trawlers, and illustrate a subtle but significant response to and dependence on man's activities.

In summary, the omnipresence of man, often in high densities or with widespread impacts, has had profound and lasting influences on most ecosystems. Even some of the more minor impacts of man can generate negative and subtle changes that give new direction to systems. This has resulted in the establishment through time of certain organisms and systems that are dependent upon man-caused perturbations. As a result, any sophisticated system analysis must include the direct and indirect effects of man [Odum et al. 1979].

Perturbations, in turn, appear to have profound influences on man. The cyclonic belt that sweeps across North America and Europe brings uncertain, changeable weather with abrupt climatic extremes common. Some of the most vigorous civilizations in the world have developed under this regime, perhaps as a result of the human stimulus and stress created by the frequent, unpredictable weather changes. The sharp and abrupt juxtaposition of storms, variable air pressures, temperatures, winds, ionization and moisture in these regions may be invigorating and fortuitous to the humans living there. This may be related to the fact that human systems remain healthy when they are periodically subject to exertion and stimulation just as other systems must be repeatedly perturbed.

PERTURBATION SERIES

Many systems are not dependent upon or limited by a single type of perturbation. They are instead controlled by a number of different kinds of perturbations occurring in sequence. Each perturbation often appears to be a random event unrelated to other forces, and of little consequence to the system. But when the impacts of these various kinds of disturbance factors are summed up through time, positive relationships emerge.

Most grasslands, for example, are the product of a series of perturbations that occur in a more or less random order. Such things as downpours, flooding, droughts, lightning, fires, hailstorms, summer frosts, herbivore buildups, insect outbreaks, ice storms, snow breakage, high winds, tornados, blizzards, ice blasting and winter desiccation occurring frequently and repeatedly favor grasslands while deterring brush and forest. Many of these individual events appear to be only minor disturbances or, at best, intense climatic conditions that are ineffective as controlling factors. But when they are connected to other, seemingly minor disturbances and isolated events, patterns emerge that show that their cumulative effects are responsible for the production and maintenance of certain perturbation-dependent systems. In some systems the precise sequence of occurrence of the various perturbations is important as in fire-flood sequences. But in other cases, although two or more factors may be linked because of biotic or environmental conditions, the order of occurrence is unimportant in comparison to the cumulative effects. In some instances, two or more perturbations may occur together or in concert, and this may intensify their impact.

Although ecologists recognize obvious perturbations as limiting factors that control certain organisms and ecosystems, it may be that in reality systems are more commonly controlled and maintained by sequences or series of various perturbations.

PERTURBATION CHARACTERISTICS

The effects of disturbances on dependent organisms and ecosystems differ from those of stress and perturbation in general. In many ways, the terms "disturbance," "catastrophe" and "perturbation" are inappropriate when dealing with dependent organisms and systems. Perhaps these perturbations or disturbances should be more appropriately termed short-lived phenomena, interruptions of continuity, sudden and brief natural forces, pulse events or extreme cases of environmental or biotic heterogeneity.

The differences between pulse events that relate to dependent systems and general perturbations can be best understood by comparing their characteristics and effects. General perturbations and stress usually create instability, reduce species diversity, set back the successional development, produce declines in productivity, cause degradation of the system or result in abrupt changes in the organisms or systems. Events on which organisms and systems are dependent, however, are repetitive and have been so for some evolutionary time. Then the so-called perturbations actually become expected or anticipated events that are part of natural systems. Even though some of these determinant occurrences are not now predictable, they usually turn out to be periodic or cyclic, while general perturbations are usually unexpected and nonphasic. Some of these events have not been recognized as being repetitious and therefore predictable, because of the short-time awareness of them or because they have been only viewed or studied narrowly and not considered in a broader time perspective [Shugart 1978]. Even the events that humans consider as once-in-a-lifetime occurrences may not be exceptions to those species involved. Organisms and the communities in which they exist must be considered in terms of the species' lifespans and system cycles. Unusual events need to occur only once in a climatic or geomorphological/geological cycle to have profound effects, or once in the life of an organism to be controlling factors [Vogl 1978]. Whittaker's [1974] "cata-climax" concept, for example, shifts emphasis to the frequency of disturbance as it relates to the lifespan of the organisms and away from the traditional views of secondary succession. This approach makes it easier to recognize perturbation-dependent systems. Similarly, Harper [1977] pointed out that the vegetational composition is better understood by looking at the historical pattern of disturbance than by considering some hypothetical successional endpoint.

Perturbations in dependent systems usually help to ensure stability, often increase species diversity, maintain vegetational development and create conditions for maximum productivity and energy capture. Disturbances promote diversity and stand stratification. General perturbations and stress are considered cataclysmic events that traumatize systems, but perturbations in dependent ecosystems not only serve as driving forces, but also act as organism and system catalysts.

In perturbation-dependent systems there is often no retrogression or successional setback, but rather a progression, maintenance or cyclic succession [Vogl 1969a, 1970b]. Although succession is variable and sometimes difficult to predict, the magnitude and latitude of variation normally stays within definable limits for any system. In other words, the effects of each pulse event follow predictable paths or patterns which vary in width depending upon the system. In some systems there may be several tracks that the system takes depending on the time when the system is affected (frequency) and the intensity of the so-called disturbance. It is under these latter circumstances that the term perturbation comes the closest to one of its definitions that implies a deflection or an offset. According to Webster's

Dictionary, a perturbation in astronomy is an irregularity in the motion or orbit of a heavenly body caused by some force other than that which determines its usual path. Earth cycles are somewhat analogous to the orbits of heavenly bodies, but the more variable paths of these systems are not directly comparable to the precise paths of celestial bodies, and, in addition, the terrestrial forces do not deflect the cycles into new, unknown or unpredictable pathways.

Perturbations in dependent systems sometimes cause widespread mortality or partial destruction as do general perturbations, but differ in that such losses are only momentary or temporary. Any destructive effects are quickly countered by the stimulating effects created by the so-called disturbance. These revitalizing effects are sometimes overlooked because catastrophic events seem to attract our attention only while they are happening. It usually takes a longer look to attain the necessary perspective to assess natural phenomena properly. It is important to connect cause and effect in each case because overgeneralization about perturbations in the past has usually categorically closed further scrutiny and differentiation.

The stimulating effects of short-lived events in dependent ecosystems involve classical sigmoid recovery curves of exponential increase to create oscillations of such magnitude that they might better be described as surges (Figure 1). This is in contrast to random perturbations in independent systems in which recovery is slow [Likens et al. 1978], is often composed of different species than existed prior to the disruption, and in which post-perturbation conditions usually represent some form of degradation.

Although these surge responses are common to perturbation-dependent systems, the specific response mechanisms or physiological causes are not always completely understood. The rapid and stimulating recovery responses are related to the alignment of a multiplicity of natural forces within organisms with the occurrence of each pulse event. This results in the synchronous amassment of numerous biological processes that produce a definite surge. Perturbations such as rains, floods and fires often synchronize the phenology of plants and animals. The organisms within perturbation-dependent systems have evolved to thrive on interruption and injury by responding with spurts of energy that are augmented by the new environmental conditions created by the disturbance factors. Without these periodic perturbations that produce a wavelike pattern when plotted [Loucks 1970], these systems usually deteriorate with declines in species diversity, biomass and organism growth and vigor. In extreme cases, systems that are denied perturbations eventually become vulnerable to nature's alternatives which may result in their degradation [Vogl 1977a].

The responses of dependent plants and vegetational systems to various natural forces are often remarkable and have been recognized by both ancient and primitive peoples. Animal recovery is usually tied to these plant responses. The cutting, pruning and girdling of various woody plants for increased production of leaves, fruits and wood have their origins in antiquity. Firewood, for example, was vigorously generated on a sustained yield basis by coppicing trees. This was accomplished by periodically pruning off terminal branches which responded by producing more numerous and rapidly growing stemsprouts. Modern gardeners are also aware of the strong natural responses to shock perturbations. The more that roses are cut and pruned, for example, the more flowers that are produced. Vineyards are pruned and girdled annually to ensure maximum grape productivity.

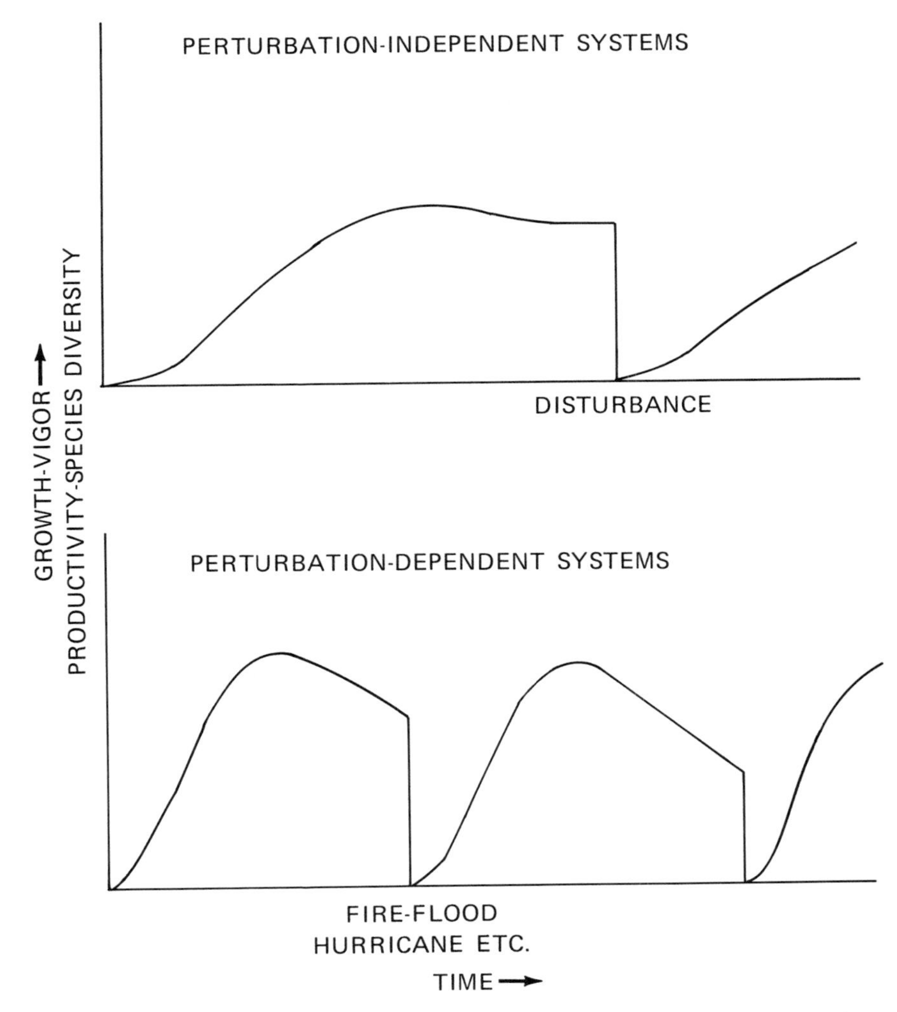

Figure 1. Disturbances in general ecosystems create vegetational setbacks and complete recovery is slow, whereas disturbances in perturbation-dependent ecosystems usually stimulate pulses of growth which rapidly decline unless disturbed again.

In many ways the strong growth response in plants is similar to inversity principle [Vogl 1978] responses in animals. The inversity principle, as recognized by wildlife biologists, states that the rate of reproductive gain of an animal species is inversely proportional to the population density at breeding time. This means that if a certain number of animals are harvested each year, the natural recovery responses will react to return the population to preharvest numbers. As a result of nature's surge responses, man has learned that such things as cutting and burning can be used as powerful forces for his benefit, as the plant removal is usually compensated by renewed growth.

A variety of natural forces are capable of amassing and invoking strong healing and self-repairing powers in plants and animals. In perturbation-dependent organisms and systems there is a reliance on these forces to produce a stimulus, boost or vitalization. But if the frequency of these forces is increased beyond their natural and normal occurrences, or beyond the response capabilities of the organisms or systems involved, degradation can take place [Vogl 1977a]. In some systems there are physical deterrents that prevent premature perturbations [Vogl 1977b], but sometimes they can occur before the recovery responses are complete and this can deplete reproductive, rehabilitation and reenergizing abilities, and adversely affect those systems.

A few additional characteristics of perturbation-dependent organisms and systems are worthy of mention. It should be remembered that these characteristics are not common to all perturbation-dependent species and systems.

Organisms of dependent systems tend to have broad amplitudes of tolerance, that is, they are adaptable and flexible, with a high degree of resiliency. The species are often opportunistic pioneers, short-lived, rapid-growing or quickly multiplying and occur in even-aged groups with few other species present. Most such species are associated with the early stages of succession because the late and climax stages exclude them and generally do not invite natural catastrophes.

Dependent systems are often characterized by periodic fluctuations and environmental extremes. They tend to have a low vulnerability to upset and rapid recovery rates [Hull and Muller 1977]. Perturbation-independent systems, conversely, are extremely vulnerable to upset, and if they are failing or unbalanced they are upset by even the slightest disturbances.

In some instances, occurrences of a perturbation in dependent systems sets up conditions or ensures the occurrence of the next perturbation. It is probable that as perturbations shift and their effects change through time, so also do the systems shift to respond and survive. In some cases, the perturbations and the ecosystems are inseparably linked with the system producing the perturbations and the perturbations shaping the system. Systems that are frequently perturbed generally have more rapid recovery rates than those subject to less frequent pulse events. Often the longer the time between perturbations, the bigger will be each perturbation.

Some perturbation-dependent systems not only decline in growth and productivity if denied exposure to the various natural forces, but also often possess little or no natural reproduction until perturbed. Some perturbation-dependent species are uncommon until the right conditions are created and then they "explode." Most dependent systems follow a tear and build sequence, with death as an essential part of each cycle or rhythm. The repetitious pulse events actually stabilize the systems [Odum 1969; Loucks 1970].

In conclusion, it is obvious that perturbation-dependent systems have certain distinct characteristics and react in predictable ways along established pathways. The term perturbation appears to be inappropriate for these dependent systems in that the systems are not disturbed or deflected but are stimulated and renewed. In addition, the remarkable recovery responses of perturbation-dependent systems are considered to be the most outstanding feature of these systems.

ACKNOWLEDGMENTS

I thank my colleague and friend Roland Case Ross for assisting in the literature search of this subject and for sharing his lifetime of study and experience with me in numerous and invaluable discussions of perturbation-dependent ecosystems. I appreciate the time spent by Drs. Harold Biswell and C. H. Muller in reviewing this manuscript and their many helpful comments and criticisms.

REFERENCES

Allen, C. T., and D. E. Foster. "Seasonal food habits of the desert termite on a shortgrass prairie in west Texas," *Res. Highlights Noxious Bush and Weed Control Range Wildlife Manage.* 8:49 (1977). Texas Tech University, Lubbock, TX.

Anderson, E. *Plants, Man and Life* (Berkeley, CA: University of California Press, 1967).

Baker, H. G. *Plants and Civilization* (Belmont, CA: Wadsworth Publication Co., 1970).

Baker, H. G. "Human influences on plant evolution," *BioScience* 21:108 (1971).

Barbour, M. G. "Is any angiosperm an obligate halophyte?," *Am. Mid. Nat.* 84:105–120 (1970).

Barrett, G. W., G. M. Van Dyne and E. P. Odum. "Stress ecology," *BioScience* 26:192–194 (1976).

Beatley, J. C. "Dependence of desert rodents on winter annuals and precipitation," *Ecology* 54:721–724 (1969).

Beatley, J. C. "Phenological events and their environmental triggers in Mojave Desert ecosystems," *Ecology* 55:856–863 (1974).

Bolton, R. B., Jr., and R. J. Vogl. "Ecological requirements of *Pseudotsuga macrocarpa* in the Santa Ana Mountains, California," *J. Forestry* 67:112–116 (1969).

Boyd, C. E. "Production, mineral accumulation and pigment concentrations in *Typha latifolia* and *Scirpus americanus*," *Ecology* 51:285–290 (1970).

Chesapeake Research Consortium, Inc. *The Effects of Tropical Storm Agnes on the Chesapeake Bay Estuarine System* (Baltimore, MD: Johns Hopkins University Press, 1977).

Chew, R. M., and A. E. Chew. "Energy relationships of the mammals of a desert shrub (*Larrea tridentata*) community," *Ecol. Monog.* 40:1–21 (1970).

Cornell, J., and J. Surowiecki. *The Pulse of the Planet* (New York: Harmony Books, Crown Publishing Co., 1972).

Ffolliott, P. F., R. E. Thill, P. C. Warren and F. R. Larson. "Animal use of ponderosa pine forest openings," *J. Wildlife Manage.* 41:782–784 (1977).

Fuchs, V., Ed. *Forces of Nature* (New York: Holt, Rinehart and Winston, 1977).

Hansen, J. E., W. Wang and A. A. Lacis. "Mount Agung eruption provides test of a global climatic perturbation," *Science* 199:1065–1068 (1978).

Harper, J. L. *Population Biology of Plants* (London: Academic Press, 1977).

Hull, J. C., and C. H. Muller. "The potential for dominance by *Stipa pulchra* in a California grassland," *Am. Mid. Nat.* 97:147–175 (1977).

Johnson, V. R. *Sierra Nevada* (Boston: Houghton Mifflin Company, 1970).

Juhren, M., F. W. Went and E. Phillips. "Ecology of desert plants. IV. Combined field and lab work on germination of annuals in the Joshua Tree National Monument, California," *Ecology* 37:318–330 (1956).

Lang, J., R. C. Heald, E. C. Stone, D. L. Dahlsten and R. Akers. "Silvicultural treatments to reduce losses to bark beetle," *California Agric.* 32(7):12–13 (1978).

Lewis, W. M., Jr. "Analysis of succession in a tropical phytoplankton community and a new measure of succession rate," *Am. Nat.* 112:401–414 (1978).

Likens, G. E., F. H. Bormann, R. S. Pierce and W. A. Reiners. "Recovery of a deforested ecosystem," *Science* 199:492–496 (1978).

Loucks, O. L. "Evolution of diversity, efficiency, and community stability," *Am. Zool.* 10:17–25 (1970).

Macdonald, K. B. "Coastal salt marsh," in *Terrestrial Vegetation of California*, M. G. Barbour and J. Major, Eds. (New York: John Wiley and Sons, Inc., 1977), pp. 263–294.

McNaughton, S. J. "Grazing as an optimization process: grass-ungulate relationships in the Serengeti," *Am. Nat.* 113:691–703 (1979).

Naveh, Z. "The evolutionary significance of fire in the Mediterranean Region," *Vegetatio* 29:199–208 (1975).

Odum, E. P. "The strategy of ecosystem development," *Science* 164:262–270 (1969).

Odum, E. P., J. T. Finn and E. H. Franz. "Perturbation theory and the subsidy-stress gradient," *BioScience* 29:349–352 (1979).

Peterman, R. M. "The ecological role of mountain pine beetle in lodgepole pine forests and the insect as a management tool," in *Mountain Pine Beetle Management in Lodgepole Pine Forests*, A. A. Berryman, R. W. Stark and G. D. Amman, Eds. (Moscow, ID: University of Idaho Press, 1978).

Shugart, H. H., Ed. *Time Series and Ecological Processes* (Philadelphia: Society for Industrial and Applied Mathematics, 1978).

Singer, F. J. "Habitat partitioning and wildfire relationships of cervids in Glacier National Park, Montana," *J. Wildlife Manage.* 43:437–444 (1979).

Soholt, L. F. "Consumption of primary production by a population of kangaroo rats (*Dipodomys merriami*) in the Mojave Desert," *Ecol. Monog.* 43:357–376 (1973).

Stebbins, R. C. "Off-road vehicles and the fragile desert," *Am. Biol. Teacher* 36:203–208, 294–304 (1974).

Stoneburner, D. L. "Evidence of hurricane influence on barrier island slash pine forests in the northern Gulf of Mexico," *Am. Mid. Nat.* 99:234–237 (1978).

Taylor, J. W. "Downburst: July 4, 1977," *Wisconsin Nat. Res.* 2(4):23–25 (1978).

Tevis, L., Jr. "A population of desert ephemerals germinated by less than one inch of rain," *Ecology* 39:688–695 (1958a).

Tevis, L., Jr. "Germination and growth of ephemerals induced by sprinkling a sandy desert," *Ecology* 39:681–688 (1958b).

Tevis, L., Jr. "Interrelations between the harvester ant (*Veromessor pexgandei (Mayr)*) and some desert ephemerals," *Ecology* 39:695–704 (1958c).

Thompson, J. N., and M. F. Willson. "Disturbance and the dispersal of fleshy fruits," *Science* 200:1161–1163 (1978).

Vernon, D. P. "Studies of a caterpillar, *Hyles* (*Celerio*) Lineata (Fabricius): Sphingidae, feeding on a desert annual plant: *Abronia pogonantha*," MA Thesis, Biology Department, California State University, Los Angeles, 1975.

Vogl, R. J. "Vegetational history of Crex Meadows, a prairie savanna in northwestern Wisconsin," *Am. Mid. Nat.* 72:157–175 (1964).

Vogl, R. J. "One-hundred and thirty years of plant succession in a southeastern Wisconsin lowland," *Ecology* 50:248–255 (1969a).

Vogl, R. J. "The role of fire in the evolution of the Hawaiian flora and vegetation," Proceedings Annual Tall Timbers Fire Ecology Conference No. 9, Tallahassee, FL (1969b), pp. 5–60.

Vogl, R. J. "Fire and the northern Wisconsin pine barrens," Proceedings Annual Tall Timbers Fire Ecology Conference No. 10, Tallahassee, FL (1970a), pp. 175–209.

Vogl, R. J. "Fire and plant succession," in *Symposium on The Role of Fire in the Intermountain West* (Missoula, MT: Intermountain Fire Research Council, 1970b), pp. 65–75.

Vogl, R. J. "Effects of fire on southeastern grasslands," Proceedings Annual Tall Timbers Fire Ecology Conference No. 12, Tallahassee, FL (1972), pp. 175–198.

Vogl, R. J. "Ecology of knobcone pine in the Santa Ana Mountains, California," *Ecol. Monog.* 43:125–143 (1973).

Vogl, R. J. "Effects of fire on grasslands," in *Fire and Ecosystems*, T. T. Kozlowski and C. E. Ahlgren, Eds. (New York: Academic Press, Inc., 1974), pp. 139–194.

Vogl, R. J. "Fire frequency and site degradation," in *Symposium on Environmental Consequences of Fire and Fuel Management in Mediterranean Ecosystems*, Palo Alto, CA (1977a), pp. 193–201.

Vogl, R. J. "Fire: A destructive menace or a natural process?," in *Recovery and Restoration of Damaged Ecosystems*, J. Cairns, Jr., K. L. Dickson and E. E. Herricks, Eds. (Charlottesville, VA: University Press of Virginia, 1977b), pp. 261–289.

Vogl, R. J. *A Primer of Ecological Principles* (Cypress, CA: Pyro Unlimited, 1978).

Vogl, R. J., W. P. Armstrong, K. L. White and K. L. Cole. "The closed-cone pines and cypresses," in *Terrestrial Vegetation of California*, M. G. Barbour and J. Major, Eds. (New York: John Wiley and Sons, Inc., 1977), pp. 295–358.

Vogl, R. J., and L. T. McHargue. "Vegetation of California fan palm oases on the San Andreas Fault," *Ecology* 47:532–540 (1966).

Ward, P. "Fire in relation to waterfowl habitat of the delta marshes," Proceedings Annual Tall Timbers Fire Ecology Conference No. 8, Tallahassee, FL (1968), pp. 254–267.

Ward, R. C. *Floods: A Geographical Perspective* (Somerset, NJ: Halstead Press, 1978).

Went, F. W. "Ecology of desert plants, I. Observation on germination in the Joshua Tree National Monument, California," *Ecology* 29:242–253 (1948).

Went, F. W. "Ecology of desert plants. II. The effect of rain and temperature on germination and growth," *Ecology* 30:1–13 (1949).

Whittaker, R. H. "Climax concepts and recognition," in *Vegetation Dynamics, Part VIII of the Handbook of Vegetation Science*, R. Knapp, Ed. (The Hague, Netherlands: Junk, 1974), pp. 137–154.

Wilson, D. E., and S. M. Hirst. "Ecology and factors limiting Roan and Sable Antelope populations in South Africa," Wildlife Monographs No. 54 (1978), 111 pp.

Young, J. A., R. A. Evans and D. L. Neal. "Treatment of curleaf *Cerocarpus* seeds to enhance germination," *J. Wildlife Manage.* 42:614–620 (1978).

CHAPTER 3

TO REHABILITATE AND RESTORE GREAT LAKES ECOSYSTEMS

John J. Magnuson
University of Wisconsin
Madison, Wisconsin 53706

Henry A. Regier
University of Toronto
Toronto, Ontario

W. John Christie
Ontario Ministry of Natural Resources
Picton, Ontario

William C. Sonzogni
Great Lakes Basin Commission
Ann Arbor, Michigan 48109

INTRODUCTION

The purpose of this chapter is to assess whether it is now feasible to rehabilitate and restore Great Lakes ecosystems. The sequence of changes that have occurred in a Great Lakes ecosystem are depicted (Figure 1) as a curved trajectory starting from an initial state or states about 300 years ago and ending at the present. Although simple, the figure is sufficiently realistic to provide approximations to the definitions of the terms "restoration," "rehabilitation," "enhancement" and "degradation." Restoration as a policy takes the ecosystem back in a rather direct route toward its initial state. Presumably, undesirable features of the initial natural state would be accepted as part of the overall package. Further degradation, more or less consistent with the degradative processes of the past two centuries, leads in

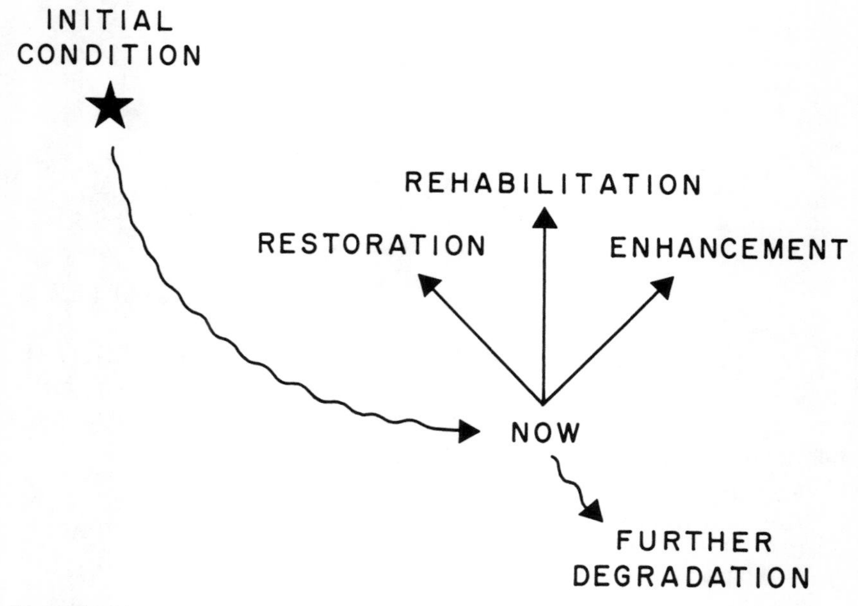

Figure 1. Diagram to illustrate the meanings of several policy options for management of natural ecosystems.

the opposite direction from restoration. Enhancement or improving the current state of an ecosystem without reference to its initial state might lead an ecosystem further from its initial state, perhaps by adding desirable manmade features and suppressing undesirable natural features. Rehabilitation may be defined as a pragmatic mix of nondegradation, enhancement and restoration. To the extent that natural ecosystem recovery can be fostered, restoration of some desirable features can be expected to be a cost-effective tactic within such a mix.

HISTORICAL PERSPECTIVE

Is restoration and rehabilitation of Great Lakes ecosystems feasible? Despite much difficult and costly work by many experts, the evidence available in 1977 did not demonstrate unequivocally that the overall quality of Great Lakes water and their aquatic ecosystems had improved. The dominant approach after the 1972 Canada-U.S. Water Quality Agreement was to combat dangerous and bothersome pollutants almost on a one-by-one basis. Each pollutant presented a challenge, the number of pollutants was quite high, and more were emerging each year. Meanwhile, fishery agencies, coordinated by the Great Lakes Fishery Commission, were expanding their efforts to control sea lamprey and to restore lake trout and other valued fisheries. Tacitly, fishery managers assumed that their colleagues in the environmental agencies, coordinated by the International Joint Commission, would succeed in containing and reversing the flood of pollution and other stresses.

These and other efforts seemed to converge on a major common challenge for environmental and fishery agencies—to rehabilitate Great Lakes ecosystems. A need was evident to clarify opportunities and options for rehabilitation and to provide a basis for much broader collaboration among the many agencies involved. The proposal for a feasibility study was first published in September 1977 by Francis and Regier [1977] and was funded in the same year as a two-year feasibility study by the Great Lakes Fishery Commission. The steering committee now consists of H. A. Regier and J. J. Magnuson as coconveners, with A. M. Beeton, W. J. Christie, G. R. Francis and M. H. Patriarche as members. This chapter is taken from the first year results of this feasibility study.

Table I in part reflects the study's approach and lists the contributors. In any restoration it is mandatory to backcast the conditions of the ecosystems prior to degradation. This image provides a target for restoration and a component of rehabilitation. Also required is knowledge of the present conditions and, even more importantly, how the ecosystems changed from the initial state to the existing state. What caused these changes? Rehabilitation and restoration are not new to the Great Lakes. How successful have past efforts been? Similarly, what techniques have worked in other aquatic rehabilitation projects that could be applied to the Great Lakes? What would be feasible in the Great Lakes? How will progress likely be influenced by future developments through processes and events not under our control? Since rehabilitation deals with social values and is implemented by institutions these elements were also considered. Only a few of these topics are discussed briefly in this synopsis of the work to date.

The existing image of the initial state is imperfect. Even in areas such as cartography where our information was relatively good 200 years ago, it is clear from an examination of old maps that the Great Lakes cannot be characterized precisely from past records, yet we have attempted to do so. Often our earliest record must serve as the benchmark. We will not detail these conditions here, but basically the lakes provided cool and cold water habitats of low fertility. Primary productivity was low, oxygen was present in deep waters all year. Even in midsummer, streamflow was abundant with cool, clear water. Wetlands, marshes, macrophytes, bays and rivers provided a rich mosaic of shallow water habitats. Large organisms dominated many taxa—perhaps as much as 50% of the total biomass of all fish was contributed by individuals over 5 kg in weight. There were numerous, locally adapted stocks of salmonines and coregonines. There were no Pacific salmon, brown trout, rainbow trout, rainbow smelt, alewife, carp, goldfish, white perch or the zooplankter *Eurythemora affinis*, and sea lamprey were confined to Lake Ontario. Some unique species or subspecies, now absent, were present, i.e., several coregonines, Michigan grayling and blue pike.

Table I. Table of Contents of First Year's Report on the Feasibility of Rehabilitating Great Lakes Ecosystems, Indicating the Approach and the Contributor

1. INTRODUCTION
   H. A. Regier[a]

2. PAST LAKE AND STREAM CONDITIONS
   H. A. Regier[a] and T. H. Whillans[a]

3. PRESENT LAKE CONDITIONS AND WHAT CAUSED THEM
   W. J. Christie[b] and N. M. Burns[g]

4. SOCIO-ECONOMIC ASPECTS OF REHABILITATION AND RESTORATION
   R. C. Bishop,[d] D. R. Talhelm[l] and J. H. Donnan[m]

5. APPRAISAL OF PAST AND PRESENT REHABILITATION AND RESTORATION INITIATIVES
   M. H. Patriarche[h]

6. EVALUATION OF REHABILITATION AND RESTORATION APPLIED TO OTHER ECOSYSTEMS
   S. M. Born,[d] J. Cairns, Jr.,[e] C. Goldman[i] and J. J. Magnuson[d]

7. TO WHAT EXTENT ARE REHABILITATION AND RESTORATION OF HABITATS AND COMMUNITIES POSSIBLE?
   W. C. Sonzogni,[c] S. C. Chapra,[k] D. E. Armstrong,[d] D. Weininger,[d] R. M. Horrall,[d] J. J. Magnuson[d] and H. A. Regier[a]

8. IMPACT OF FUTURE DEVELOPMENTS ON REHABILITATION AND RESTORATION STRATEGY
   M. G. Johnson[g]

9. INSTITUTIONAL ARRANGEMENT FOR REHABILITATION AND RESTORATION
   J. W. Bulkley[j] and G. R. Francis[f]

10. WHAT SHOULD WE DO NEXT?
    H. A. Regier[a]

[a]University of Toronto–Toronto, Ontario, Canada
[b]Ontario Ministry of Natural Resources–Picton, Ontario, Canada
[c]Great Lakes Basin Commission–Ann Arbor, Michigan, USA
[d]University of Wisconsin–Madison, Wisconsin, USA
[e]Virginia Polytechnic Institute–Blacksburg, Virginia, USA
[f]University of Waterloo–Waterloo, Ontario, Canada
[g]Canada Centre of Inland Waters–Burlington, Ontario, Canada
[h]Michigan Department of Natural Resources–Ann Arbor, Michigan, USA
[i]University of California–Davis, California, USA
[j]University of Michigan–Ann Arbor, Michigan, USA
[k]Great Lakes Environmental Research Lab–Ann Arbor, Michigan, USA
[l]Department of Fisheries and Wildlife, Michigan State University, East Lansing, Michigan, USA
[m]Environmental Approvals Branch–Toronto, Ontario, Canada

The degree of degradation of Great Lakes ecosystems is generally correlated with the population density in the area [Beeton 1969; Christie 1974]. However, it is necessary to be more specific about causality to effect a management action. Major changes to Great Lakes ecosystems have resulted from direct and indirect modifications of the habitat and biotic communities.

PHOSPHORUS LOADINGS

The lakes differ greatly in the observed changes in concentration of dissolved ions (Figure 2). Lake Superior, with a large area and a small, relatively

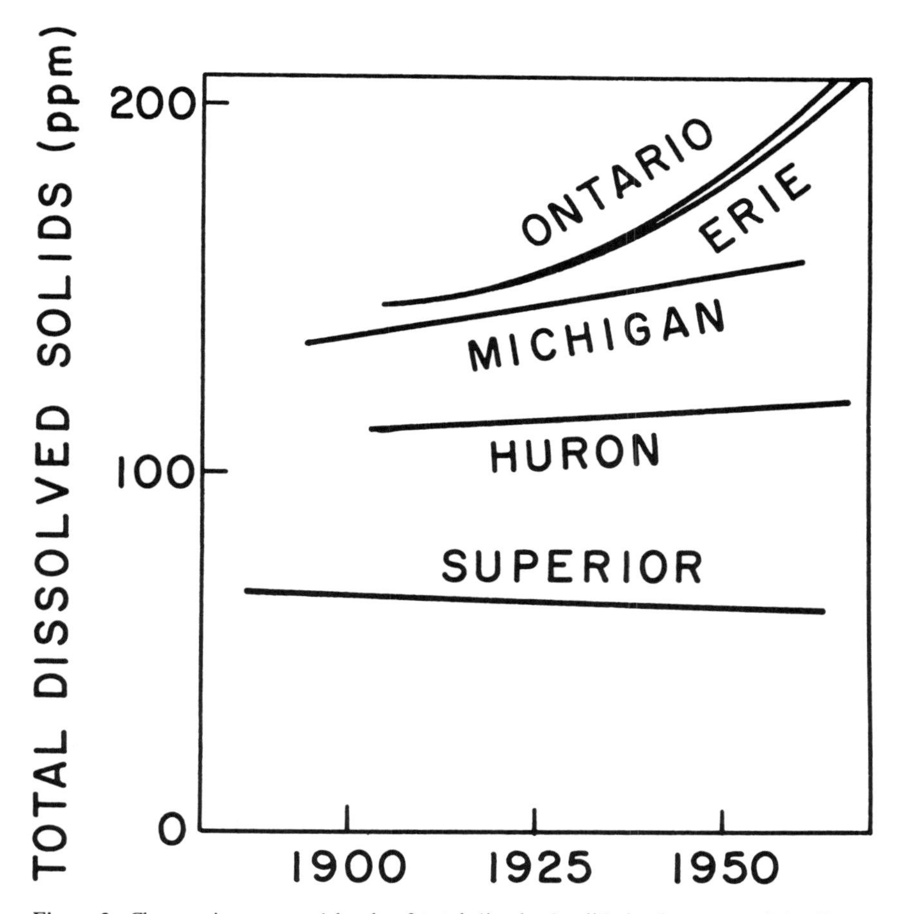

Figure 2. Changes in measured levels of total dissolved solids in the water of the Great Lakes. (Modified from *Great Lakes Basin Framework Study*, Appendix 8 Great Lakes Basin Commission, 1975.)

undeveloped forested drainage basin, has the lowest concentration of total dissolved solids (TDS), which even appears to have been declining slowly in recent years. Both Lake Erie and Lake Ontario have experienced large increases in TDS and have been negatively influenced by eutrophication and the accumulation of chemical pollutants. The increase in phosphorus loading to Lake Erie has been roughly proportional to the population increase in the basin: from approximately 3 million in 1900 to 13 million in 1975.

The increased loading of plant nutrients to Lake Erie has altered the biota. Phytoplankton biomass has increased in the Central Basin [Davis 1964]. Consequently, more organic matter settles to the bottom and increases the rate of oxygen depletion by decomposers. Surveys during 1973 and 1974 revealed that more than two thirds of the Central Basin had no oxygen in its bottom waters [Water Quality Board 1975]. Zooplankton assemblages in the lower Great Lakes (Lake Erie and Lake Ontario) now contain more forms tolerant of eutrophic conditions than in the past. Open water zooplankton communities are still "healthy," while nearshore communities are significantly altered near the large cities and in the highly eutropic inner bays [McNaught 1975]. The benthic animal *Stylodrilus heringianus*, an oligochaete of oligotrophic lakes, is conspicuous by its absence in Lake Ontario near large cities [Nalepa and Thomas 1978].

With increased eutrophication and human activity, beaches collect the flotsam of civilization, and the alga *Cladophora*, for example, becomes an especially noxious impediment to recreational use when it washes ashore and rots. Many miles of beaches in the lower lakes are lost to recreational use each year on this account [Neil 1976]. Since 1958 essentially all suitable bottom in Lake Ontario supports a lush growth of *Cladophora* during the productive season. Shoreline accumulations sometimes reach 50 feet wide by 2 feet deep [Neil 1976]. *Cladophora* problems in the lower lakes have no doubt affected shoreline property values as well as having fouled public beaches. Only Lake Superior is essentially free of *Cladophora* problems.

Phosphorus loadings to the Great Lakes for 1976 are summarized in Table II. Lake Erie receives by far the greatest proportion of the phosphorus input to the Great Lakes. Lake Huron receives the least, although Saginaw Bay receives much of this load and as a result shows severe signs of eutrophication. Phosphorus from municipal wastes is the most important source, particularly in terms of the potential for rehabilitation. Runoff from land is also an important source of phosphorus, but, unlike point sources, it appears based on recent information that less than half of this is in a form usable for plant growth [Armstrong et al. 1978; Logan 1978]. Phosphorus also originates by natural processes such as shoreline erosion. However, much of this phosphorus is also in a form not available to support plant growth. Shoreline erosion is a good example of a natural process that has been going on for centuries. Atmospheric fallout of phosphorus is another important source. To large, unproductive waters such as those of Lake Superior, atmospheric contributions may actually be beneficial.

Recommendations of the International Joint Commission Technical

Table II. Total Phosphorus Loading to the Great Lakes in 1976. Total loading was estimated at about 57,000 metric tons per year. The percent to each lake and from various sources for the upper and lower lakes is given. Modified International Joint Commission Technical Group to Review Phosphorus Loadings (1978) and the International Joint Commission's Pollution from Land Use Activities Reference Group (1978).

Lake	Percentage to each lake
Erie	44
Michigan	18
Ontario	17
Superior	13
Huron	8

Sources	Percentage from each source	
	Upper Lakes	Lower Lakes
Point	17	32
Tributary (nonpoint)	29	37
Shore erosion	37	27
Atmospheric	17	4

Group on Phosphorus Loading suggests that phosphorus limitations are technically feasible to achieve nondegradation. For example, limiting the discharge of municipal waste treatment plants to 1 mg/l of phosphorus would be sufficient to achieve nondegradation of Lakes Michigan, Huron and Superior. In addition to reducing the discharge from municipalities, a 50% reduction in diffuse sources would be required for Lakes Erie and Ontario to reach a desired "target" phosphorus load. The latter would be much more difficult to accomplish but clearly points out the benefits and importance of managing the tributaries and their watersheds as part of a rehabilitation program. For Lake Erie it is projected by the Technical Group that the "target" loading would result in 90% of the bottom waters of the Central Basin remaining oxygenated. Such action would have the additional benefit of preventing substantial phosphorus release from the sediments. Perhaps the most surprising element of the projections of the International Joint Commission Technical Group on Phosphorus Loading 1978 is the rapid response time of the ecosystems to reduced phosphorus loading–1–2 years for Lake Erie; 5–7 years for Lakes Michigan, Huron and Ontario; and 15–20 years for Lake Superior, which is presently in the best condition.

MICROCONTAMINANTS

Microcontaminants pose a major problem in the rehabilitation of Great Lakes ecosystems. The chlorinated hydrocarbons such as PCBs, DDT and

Mirex are especially problematic because of their stability in the environment and their accumulation by fish. The latter makes the fish unfit for human consumption and consequently creates financial hardships for the fishers. The contaminants appear to pose less direct problems for the fish themselves. In fact, when the Lake St. Clair walleye fishery was closed in 1970 owing to mercury contamination, the lowered exploitation resulted in a rapid recovery of the population which had been overfished [Great Lakes Fishery Commission 1974]. In addition to the above compounds, other microcontaminants such as chlorophenols and trace metals may cause problems in the Great Lakes.

The rate of change of a microcontaminant level in the water column and its biota is determined by:

1. the input from external sources;
2. losses from biodegradation, surface water discharge, volatilization, harvesting of biota containing microcontaminants and sedimentation; and
3. recycling from bottom sediments.

Weininger (Table I) evaluated the persistence of PCBs in Lake Michigan fish as part of this feasibility study. His projections are based in part on the behavior of DDT after its use was banned in 1970 (Figure 3). DDT levels in coho salmon in Lake Michigan decreased rapidly after the ban to about 18% of its maximum level in 1970. Weininger explains this by considering two major sources of the contaminant to the coho salmon—prey that derive food from the pelagic community and prey that derive food from the benthic community. He hypothesizes that the rapid decrease reflects the removal of DDT from the water column. The second part reflects benthic transport. Since the rate of sedimentation is low, the microcontaminants in the surface sediment remain available to biota for recycling into the food web. Residence time of DDT in the water column appears to be less than two years.

Lake trout in Lake Michigan contain 15–35 ppm of PCBs [Willford 1977; Veith 1975; Weininger 1978]. Calculations by Weininger suggest that less than 1 ppm of this could originate from direct uptake from the water which contains 10 ng/l. How rapidly the PCB levels will decline in lake trout after PCB sources to Lake Michigan are eliminated will depend on the conversion efficiency of their prey and the proportion of their prey derived from the pelagic pathway. Levels should decline most rapidly if they eat prey with high conversion efficiency that derive their food from the pelagic environment. PCB levels in lake trout also should decline rapidly once PCB inputs are halted if the pelagic pathway is dominant (for DDT in coho salmon, about 18% is thought to be derived from the benthic pathway). However, because PCBs were introduced into the environment accidentally, diffuse PCB sources such as landfills will probably continue to release PCBs for many years. Nevertheless, some point sources can be stopped by discharge elimination and perhaps by dredging [Hetling et al. 1978], which should at least reduce PCB levels in the future.

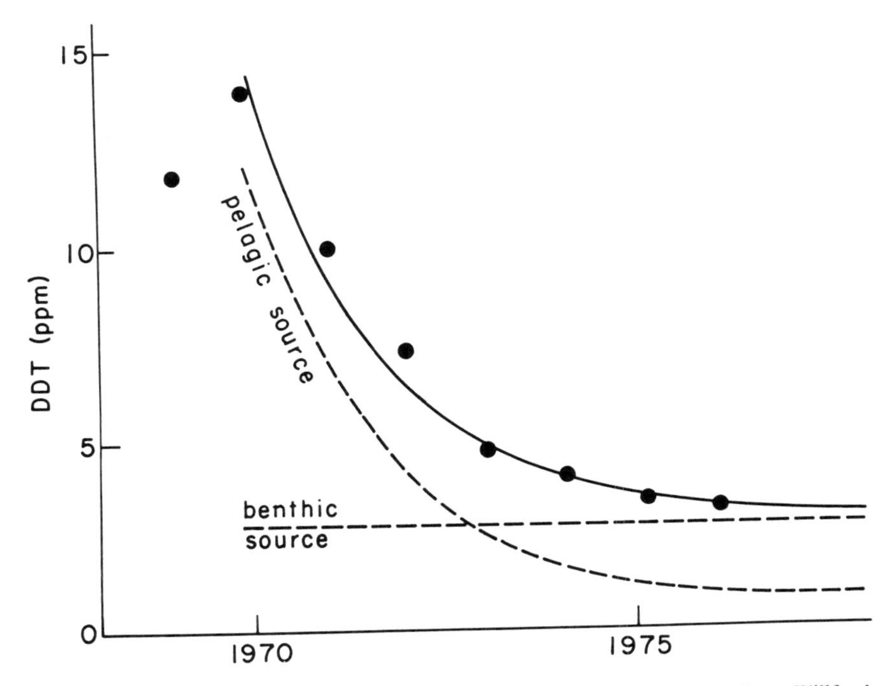

Figure 3. DDT concentration observed in Lake Michigan coho salmon from Willford [1977]. Dashed lines show the decreasing pelagic and constant benthic components based on a 1970 DDT input cessation. Interpretation is by Weininger and Armstrong (Table I).

Experience with DDT showed good public support for effective regulation of the input of synthetic microcontaminants. Prospects for rapid recovery of the ecosystem following control of inputs appear more optimistic (Figure 3) than we originally had reason to expect. The consequences of these substances to human health are so serious, that control will most likely be enthusiastically supported by society. Consequently, in the long run, some of the problems with synthetic microcontaminants may actually be resolved sooner than those of more commonplace industrial and domestic wastes.

PHYSICAL MODIFICATION OF HABITAT

Physical modification of habitat has occurred in the rivers, harbors, bays and inshore zone. Rehabilitation of physical habitat is important to reestablish the complex mosaic of inshore experiments.

On the Ganaraska River flowing into Lake Ontario, mills and dams reached maximum development about 1870 and have been declining steadily ever since (Figure 4). Many dams have fallen into disuse and a recently constructed fish pass in the last major dam at the city of Port Hope is a good example of small-scale rehabilitation.

Rivers tend to be most affected near the mouths. Brook trout are seldom seen in the lower reaches of rivers in which they were once plentiful [Hallam 1959] because of warming effects. The tendency to build large cities at the mouths of rivers has had serious impacts. The Cuyahoga River at Cleveland is an obvious example. A great deal of money is being spent upgrading the unacceptable effluents from the industries there, but no appreciable improvements to the Cuyahoga can be observed yet in the biota [Great Lakes Water Quality Board 1976]. The polluted lower reaches of rivers such as the Cuyahoga severely limit the habitat which the rivers could afford to migratory fishes.

Harbors are typically also river mouths, so they extend pollution influences well into the lakes. Pollution aside, shoreline modifications and dredging have profound effects and in many cases fish and wildlife habitat may never be recovered. The early Toronto harbor contained extensive

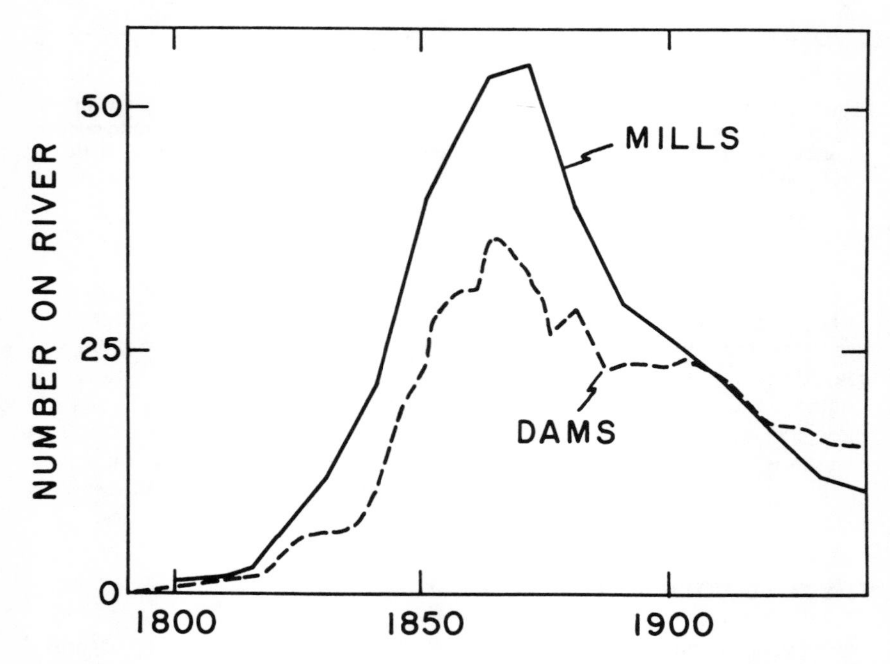

Figure 4. Number of mills and dams on the Ganaraska River which flows into Lake Ontario on the north shore. (Modified from Richardson 1944.)

marshes and beds of macrophytes which have been largely eliminated by dredging, siltation and shoreworks [Whillans 1977]. Early fishing conditions described by Whillans were excellent. Today the harbor proper no longer supports fishing.

In the last century dredging for stone or "stone hooking" was an extensive practice in Lake Ontario [Whillans 1977]. The practice extended over 75 years and employed as many as 40–50 ships along the northwest shoreline alone. The suspicion is that cobbly patches such as those used by whitefish and lake trout for spawning were prime targets for these operations and that this could have had serious impacts on fish recruitment.

Another modification of the physical habitat worth mentioning is the use of inshore water for cooling in connection with electric power generation. Entrainment and impingement losses are probably more important than the heating of the waters itself. Many native fish species in the Great Lakes have pelagic larvae and potential mortalities from entrainment need to be estimated and evaluated as impediments to restoration of fish stocks. Alternatives to minimize impact should be tested such as deepwater intakes and discharge. On the positive side, lake trout have been using some of the shoreworks of power plants in northern Lake Michigan as spawning sites. This suggests that replacing cobble previously mined from Lake Ontario could enhance reproduction of trout and whitefish. Also, because shoreworks will continue to be built, attention should be given to designing them to the ecological advantages of lake fauna.

Perhaps the key to success of any plans to rehabilitate the coastal zone is to stimulate people of nearby settlements to take a personal interest in these shoreline ecosystems. Coastal areas can be scenic spots in which people take pleasure and pride, as many early paintings and writings attest.

MODIFICATION OF BIOTIC COMMUNITIES

In addition to habitat modification, direct and indirect modifications of the biotic communities have also been important. We have already mentioned that the fish taxa were probably represented by large individuals with as many as 50% of the biomass being of fish over 5 kg. Today, the proportion of fish greater than 5 kg is infinitesimally small in the Great Lakes. The original large fish perhaps played roles not unlike the roles of the large trees in a mature forest. Compared to other organisms of the lake ecosystems, many fish were relatively massive and old, in some cases (lake sturgeon) reaching ages of over 50 years. Thus, they must have had a "conservative" function analogous to that of trees on land.

Significant changes have occurred in the fish stocks of all the Great Lakes. As with other elements of the biota, Lake Superior has had the least drastic changes. Lake Ontario was probably the earliest to be affected, followed by Lake Erie, Lake Michigan and Lake Huron, which has been affected only recently. But even in Lake Superior the effects of man are apparent. By the late 1960s catches had dropped to 37% by weight and 31% by value compared to the early 1940s [Great Lakes Basin Commission 1975]. Much of this decline was from the dramatic declines in catches of valued species such as lake trout and lake herring. Even earlier the accumulated capital of the old fish had been gradually exploited (mined), as was done with large trees on land. Thereafter, fisheries had to make do with the annual interest on a much diminished though much more productive capital stock. In the classical manner of a common property resource of limited scope, the gradually intensifying fisheries soon led to overfishing of the preferred stocks. Like many predators, the fishers switched to progressively lower valued species as prey. The capital stock of high valued fish species in the Great Lakes is now a small fraction—perhaps 5%—of what was in the Great Lakes two centuries ago.

The sea lamprey has been the most widely publicized cause of Great Lakes stock problems [Smith 1968; Lawrie 1970; Christie 1974]. Penetrating to the upper lakes from Lake Ontario where it had been native [Lark 1973], it brought about almost a complete collapse of the lake trout stocks in the upper three lakes by direct predation on the trout. In Lake Michigan lake trout was extinct by 1956. The lamprey invasion resulted from the construction of the Welland Canal and was a direct result of the westward expansion of shipping to the Great Lakes region. While overfishing has seldom been clearly indicated in the collapse of Great Lakes fish stocks, it is believed to have exacerbated the effects of the lamprey. However, the losses of lake sturgeon and some stocks of whitefish were documented for the last century, before such factors as eutrophication and invasion or introduction of pest species like the lamprey became problems [Christie 1974].

Introduction and proliferation of exotic pest fishes like the alewife and the rainbow smelt have also shared the blame for stock losses [Smith 1970; Lawrie 1970]. The drastic impact of the alewife on virtually all fish species in Lake Michigan was documented by Wells and McLain [1972]. Christie [1974] suggested that the proliferation was possible because of the virtual absence of predators in the lakes; populations of the lake trout and burbot collapsed because of sea lamprey invasions. Dieoffs of the abundant alewife at the peak of their abundance [Colby 1971] endangered water supplies and marred beaches for recreational uses.

Perhaps the best examples of restoration of Great Lakes ecosystems came out of the events that followed sea lamprey and alewife invasions. These efforts were coordinated by the Great Lakes Fishery Commission. Initially weirs were constructed to prevent the spawning of sea lamprey in tributaries to the upper Great Lakes. Later, a toxin called 3-trifloromethyl-4-nitrophenol (TFM) selective for ammocoetes was applied in streams where the young spent up to 5 years [Applegate et al. 1961]. First treatments were to Lake Superior because the lake trout population there was still present although in reduced numbers. The program was successful and extended to Lake Michigan and Lake Huron. In addition, young lake trout were stocked to re-establish naturally reproducing populations. The lake trout survived and grew, feeding primarily on alewife and smelt. At present there are large numbers of lake trout in the upper Great Lakes but to date almost no natural reproduction of introduced lake trout has been observed. Serious attempts are now underway by Horrall (Table 1) and others to determine why the once native species, the lake trout, is so difficult to reestablish. Hypotheses being considered include:

1. contaminants such as PCBs and DDT reduce viability of eggs and fry;
2. inappropriate stocking methods result in failure of fish to find suitable spawning areas;
3. genetic makeup of stocked lake trout is inappropriate for Great Lakes habitats;
4. age distributions or abundances of stocked lake trout are inappropriate;
5. the now abundant alewife and smelt interfere with survival of young trout;
6. habitat changes have degraded spawning reefs.

In 1978 a total of 5.2 million young lake trout were scheduled for stocking in the Great Lakes. This expensive venture was once done free of charge by nature.

In further efforts to rehabilitate fish communities, the state of Michigan took the initiative in enhancement by stocking nonnative salmonines to the lakes–namely, coho salmon, chinook salmon, steelhead trout, brown trout and Atlantic salmon. The purposes were both to crop the overabundant alewife and to provide a high-value sport fishery in the region. The program was highly successful [Borgeson and Tody 1967; Borgeson 1970; Rybicki 1973].

Now the question has arisen as to how extensive a predator planting program should be, so as not to overexploit the forage base of alewives and smelt. Lake trout and coho salmon have fed primarily on the exotics–rainbow smelt and alewife. Brown [personal communication] estimates as a first approximation that perhaps as much as 90,000 metric tons of alewife were eaten per year or about 1/3 of the stock in Lake Michigan in certain years. This predation pressure may have reduced the abundance of alewife. Moreover, some apparent recoveries in populations of yellow perch may be in part due to the decline in alewife.

Further reductions in alewife might result from even more extensive planting of predators, but if the resurgence of the native coregonines and percids is to be achieved, a selective removal of alewife and rainbow smelt might be encouraged by use of intensive and highly selective fishing efforts. Fishes in the family to which alewife belongs, the Clupeidae, have routinely been overexploited by man [Murphy 1977]. This suggests that alewife can be overfished as well. The clue to selective fishing is to know enough about how a species is distributed in comparison with others so that, for example, aimed trawling can be implemented intensively and precisely on a particular species. Since the distribution of alewife is closely related to bottom temperature, it may be feasible to haul a trawl along a particular isotherm in an attempt to fish selectively. Example data of this sort were collected by Brandt [1978] (Figure 5). Fishing at 13–14°C during the day produced catches almost exclusively of adult alewife and ninespined stickleback. By putting rollers on

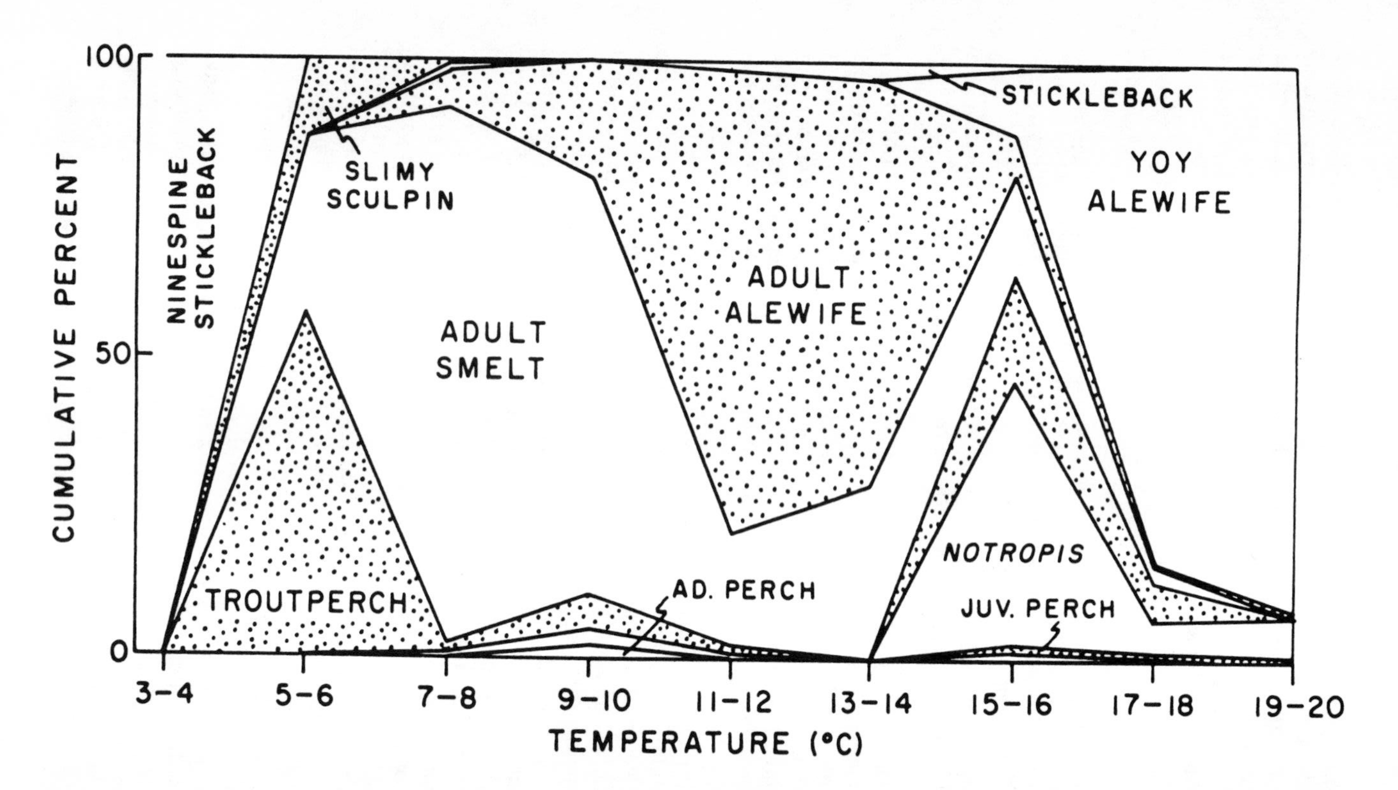

Figure 5. Distribution of fishes caught in bottom trawls in relation to bottom water temperature off Grand Haven in Lake Michigan during day in September 7–13, 1977 [Brandt 1978]. Note that both alewife and smelt are exotic species and dominate the catches. Percentages based on geometric means of catch per unit of effort at each depth.

the net or adjusting mesh size, the catch of stickleback might be eliminated. Alewife also migrate off bottom at night into areas near the thermocline. Large midwater trawls might be used at this time to avoid more bottom-oriented species. Regardless of method, selective reduction in alewife and smelt would be expected to produce positive responses in the native stocks of coregonines and percids. Such biomanipulation would be expected to have repercussions throughout the ecosystem. The zooplankton community, the phytoplankton community, and even the phosphorus levels would be influenced, based on experimental biomanipulations by Shapiro et al. [1975] and experience with alewife in smaller lakes [Brooks and Dodson 1965].

CONCLUSIONS

We have reviewed some of the factors causing change in Great Lakes ecosystems and have suggested ways to rehabilitate the ecosystems. Man has influenced and degraded both the habitat and the biotic communities. He has had some piecemeal successes in rehabilitation. Technical approaches to rehabilitation are outlined in Table III for chemical characteristics, physical habitat and biotic communities.

Restoration of many of the early qualities of the Great Lakes is a desirable goal, yet we will often have to settle for a partial restoration or rehabilitation. Human population in the region continues to grow. Competing uses of the waters and shorelines intensify. Our choices are difficult for society. But even with the most noble goals and enthusiastic support, restoration of the ecosystems is often impossible. This is not just a matter of impracticability—there are certain potential goals which are now lost options or even undesired options. For example, a number of new species have been introduced or have invaded the Great Lakes. We might regulate their abundance but cannot eliminate them altogether. Contemplate a program to cause the extinction of alewife or the zooplankter *Eurythemora affinis*. Also, a number of species

Table III. An Outline of Rehabilitation Approaches for the Great Lakes

I. Chemical (nutrients and contaminants)
   A. Reduce inputs of chemicals to acceptable levels
   B. Plan for the duration of their persistence or the response time of the ecosystem
   C. Apply in-lake management to lessen the influence or hasten the decline

II. Physical
   A. Prevent additional detrimental modifications
   B. Physically alter selected sites to help reestablish habitat mosaic

III. Biotic Communities
   A. Prevent introduction or invasion of new species
   B. Selectively exploit unwanted species with man or other predators
   C. Reestablish breeding populations of native species
   D. Make decisions on the ecosystem level for exploitation of the biotic community

and many genetic subpopulations have disappeared from the lakes. Further, it is unlikely that society would choose to eliminate stocking of certain salmonids that were not part of the presettlement fauna of the lakes. We would not choose to let age and size structures of species such as lake trout and walleye return to unexploited conditions. Even with a concerted effort, we must settle for rehabilitation no matter how intense our desire for the original state.

We do believe that partial restoration combined with appropriate technical enhancement can feasibly result in major rehabilitation. However, great concerted efforts will be needed to achieve major rehabilitation. The scale of the program would be similar to revitalizing cities whose cores have been blighted through social neglect, or to that of rehabilitating semiarid regions devastated by extended droughts and poor husbandry.

Whether many Canadians and Americans have sufficiently intense desire for major rehabilitation of degraded ecosystems of the Great Lakes to make it happen is still open to question. To date, people have not been provided the necessary information and some reasoned options on which to base a reasoned decision. This feasibility study being developed by the Great Lakes Fishery Commission is a contribution toward that information base.

REFERENCES

Armstrong, D. E., M. A. Anderson, J. R. Perry and D. Flatness. "Availability of pollutants associated with suspended or settled river sediments which gain access to the Great Lakes," Progress Report, EPA Contract No. 68-01-4479 (1978).

Beeton, A. M. "Changes in the environment and biota of the Great Lakes," in *Eutrophication: Causes, Consequences, Correctives* (Washington, DC: National Academy of Sciences, 1969), pp. 150–187.

Borgeson, D. P., Ed. "Coho salmon status report," Michigan Department of Natural Resources, Fish Management Report 3 (1970), 31 pp.

Borgeson, D. P., and W. H. Tody. "Status report on Great Lakes Fisheries 1967," Michigan Department of Conservation, Fish Management Report 2 (1967), 35 pp.

Brandt, S. B. "Thermal ecology and abundance of alewife (*Alosa pseudoharengus*) in Lake Michigan," Ph.D. Thesis, University of Wisconsin-Madison (1978).

Brooks, J. L., and S. I. Dodson. "Predation, body size, and composition of plankton," *Science* 150:28–35 (1965).

Christie, W. J. "Changes in the fish species composition of the Great Lakes," *J. Fish. Res. Board Can.* 31:827–854 (1974).

Colby, P. J. "Alewife dieoffs: why do they occur?," *LIMNOS* 4(2):18–27 (1971).

Davis, C. C. "Evidence for eutrophication of Lake Erie from plankton records," *Limnol. Oceanog.* 3:275–283 (1964).

Francis, G. R., and H. A. Regier. "Let's rehabilitate and restore degraded ecosystems of the Great Lakes," *Int. Assoc. Great Lakes Res. Lakes Lett.* 8(3):2–9 (1977).

Great Lakes Basin Commission. "Appendix 8: Fish," *Framework Study* (1975), 290 pp.
Great Lakes Fishery Commission. "1974 Annual report" (Ann Arbor, MI: Great Lakes Fishery Commission, 1976).
Great Lakes Water Quality Board. "Great Lakes water quality 1974 annual report" (Windsor, Ontario: International Joint Commission, 1975).
Great Lakes Water Quality Board. "Great Lakes water quality 1975 annual report" (Windsor, Ontario: International Joint Commission, 1976).
Hallam, J. C. "Habitat and associated fauna of four species of fish in Ontario streams," *J. Fish. Res. Board Can.* 16:147–173 (1959).
Hetling, L., E. Horn and J. Tofflemine. "Summary of Hudson River PCB study results," Technical Paper 51 (Albany: New York State Department of Environmental Conservation, 1978), 88 pp.
Lark, J. G. "An early record of sea lamprey from Lake Ontario," *J. Fish. Res. Board Can.* 30:131–133 (1973).
Lawrie, A. H. "The sea lamprey in the Great Lakes," *Trans. Am. Fish. Soc.* 4:766–775 (1970).
Logan, T. J. "Bioavailability of sediment phosphate," Final Report to Lake Erie Wastewater Management Study. U.S. Army Corps of Engineers, Buffalo District, Buffalo, New York (1978).
McNaught, D. C. "A hypothesis to explain the succession from calanoids to cladocerans during eutrophication," *Verhandl. Int. Verein. Theoret. Angewandte Limnol.* 19:724–731 (1975).
Murphy, G. I. "Clupeoids," in *Fish Population Dynamics*, J. A. Gulland, Ed. (London: John Wiley & Sons, 1977), pp. 283–308.
Nalepa, T. F., and N. A. Thomas. "Distribution of macrobenthic species in Lake Ontario in relation to sources of pollution and sediment parameters," in *Status of the Biota of Lake Ontario during IFYGL*, N. A. Thomas and W. J. Christie, Eds. (in press).
Neil, J. H. "Distribution," in *Cladophora in the Great Lakes*, A. Shear and D. E. Konasewich, Eds. (Windsor, Ontario: IJC for the Great Lakes Research Advisory Board, 1976), pp. 1–179.
Richardson, A. H. "A report on the Ganaraska watershed," Ontario Department of Planning and Development (1944), pp. 1–253.
Rybicki, F. W. "A summary of the salmonid program (1969–1971)," in *Michigan's Great Lakes Trout and Salmon Fishery 1969–1972* (Lansing, Michigan: Department of Natural Resources, 1973), pp. 1–17.
Shapiro, J., V. Lamarra and M. Lynch. "Biomanipulation: An ecosystem approach to lake restoration," in *The Proceedings of a Symposium on Water Quality Management Through Biological Control*, P. L. Brezonik and J. L. Fox, Eds. Report #ENV 07-75-1 (Gainesville, FL: University of Florida, 1975).
Smith, S. H. "Species succession and fishery exploitation in the Great Lakes," *J. Fish. Res. Board Can.* 25(4):667–693 (1968).
Smith, S. H. "Species interations of the alewife in the Great Lakes," *Trans. Am. Fish. Soc.* 4:754–765 (1970).
Veith, G. "Baseline concentrations of polychlorinated biphenyls in Lake Michigan fish, 1971," *Pesticides Monitoring J.* 9:21 (1975).
Weininger, D. "Accumulation of PCBs by lake trout in Lake Michigan," Ph.D. Thesis, University of Wisconsin-Madison (1978).

Wells, L., and A. McLain. "Lake Michigan: effects of exploitation, introductions, and eutrophication on the salmonid community," *J. Fish. Res. Board Can.* 29:889–898 (1972).

Whillans, T. H. "Fish community transformation in three bays within the lower Great Lakes," M.Sc. Thesis, University of Toronto (1977).

Willford, W. "PCB data on Lake Michigan lake trout" (Ann Arbor, MI: Bureau of Sport Fisheries & Wildlife, 1977).

CHAPTER 4

RECOVERY PATTERNS OF RESTORED MAJOR PLANT COMMUNITIES IN THE UNITED STATES: HIGH TO LOW ALTITUDE, DESERT TO MARINE

Anitra Thorhaug
Florida International University
Miami, Florida 33199

INTRODUCTION

During Western man's habitation of the United States, but mostly during the last 200 years, plant communities of diverse types have been systematically destroyed or drastically modified. This is now occurring in the United States (as well as the rest of the world) at an accelerating pace, particularly in some fairly fragile ecologies which are now being exploited, such as Alaska and Hawaii. Opposed to these exploitative forces which destroy or profoundly alter plant communities is the vast accumulation of knowledge about the cultivation of crop plants, one of the oldest technologies known to man. Unfortunately the techniques focus only on about 50 basic crop plants and a second group of several hundred horticultural species, and the knowledge has not been incorporated into the rest of botanical and ecological research. Thus, an unfortunately large gap occurs between knowledge in agriculture (including horticulture), which is systematic cultivation of a small number of plants with products useful to man, and the disciplines of botanical ecology, which have become involved in restoration and protection of plant species.

It is ardently hoped that ever-increasing efforts will be made to bring ecologically important climax species into cultivation by botanists interested in understanding, preserving and restoring damaged plant communities.

A symposium held at Tulane University, New Orleans, Louisiana, in June 1976, "Restoration of Major Plant Communities in the U.S.A.," cosponsored by the American Institute of Botanical Sciences, the Botanical Society of America and the Ecological Society of America, was funded by the U.S.

Energy Research and Development Administration. This chapter will summarize the issues brought forward by the symposium, especially with respect to recovery patterns. The scope was to include the breadth of the restoration experience in the United States, considering the extremes from arctic areas of Alaska to subtropical Florida and tropical Caribbean; from high alpine to marine sublittoral habitats; and from desert areas through plains and forests to the coastal marshes. Examples were particularly drawn from lesser-known restoration and management efforts, so that future potential for restoration in what have been considered difficult environments could be examined. The hope was then and is now that perhaps ecologists presently only peripherally interested could focus their attention on demonstrating restoration feasibility of new plant communities. The major finding was that many types of plant communities are, with proper study, either partially or almost completely restorable after severe damage by man. This should encourage ecologists to begin studies in their own areas which eventually will lead to a low-cost and efficient restoration process.

BACKGROUND SYMPOSIA

The symposium volume [Thorhaug 1979a] is but one of several recent attempts in the United States to bring together findings from research efforts often widely dispersed or not published in the standard biological journals found in abstracting references. Often investigators accumulate data on recovery patterns of restored communities for years, then submit a report to a state or other agency which does not publish this information in the regularly reviewed biological literature. Thus, reports of long-term studies including methods and results of recovery patterns in reclamation are very difficult to find in the usual biological literature. As a consequence, special symposia have been most helpful in understanding the breadth of the restoration effort in the United States. A symposium on reclamation of drastically disturbed lands has recently been published [Schaller and Sutton 1979]. This dealt mostly with terrestrial environments and with certain kinds of mining and stripping activities. A symposium held in Blacksburg, Virginia [Cairns et al. 1977], dealt with both restoration efforts and recovery efforts for damaged ecosystems ranging from tropical forests to marine ecosystems damaged by oil spills. A symposium sponsored by the U.S. Department of the Interior [Swank, in press], dealt with restoration of coastal marshes, mangroves and seagrasses from a "how to regulate" standpoint for decision-makers within the Department of the Interior. Rates and patterns of recovery were included. *Ecology and Reclamation of Devastated Lands* [Hutnik and Davis 1973] dealt with regional occurrences of disturbed land, mostly in the southwest, and reclamation efforts with recovery in the southwest. None of these showed as wide a breadth as the symposium which has been published as individual articles in *Environmental Conservation* and which will now be reviewed with respect to recovery.

PRINCIPLES FROM THE SYMPOSIUM

The papers published in *Environmental Conservation* were:

- Cook (1976);
- Liverman (1977);
- Seneca et al. (1976);
- Teas (1977);
- Thorhaug and Austin (1976);
- Waring (1976); and
- Webber and Ives (1978).

The review of restoration efforts begins with the submarine system and moves through the intertidal, lowland and upward to the alpine. Geographically we review systems from the tropics to the tundra. Thorhaug and Austin [1976] studied restoration of the dominant subtropical-tropical seagrass, *Thalassia*. This species forms the climax community throughout much of the nearshore and back reef waters ranging from mid-Florida to Brazil. Although in damaged areas *Thalassia* had not yet naturally recovered in fifty years, restored seedlings grew to equal original seagrass abundance within five years [Thorhaug 1979b]. Other seagrass restoration methods of "sodding" and turion planting were slower and did not have high success rates. Several of the successional species of subtropical-tropical seagrasses such as *Halodule wrightii* and *Syringodium filiforme* recolonized areas much more rapidly (although other properties such as sediment compaction and animal community recolonization were not as desirable with these seagrasses) and were amenable to "sodding" methods for restoration. *Zostera marina*, which is the dominant temperate Atlantic seagrass, was not restored with as high a success rate. Recovery time may be 20–30 years, since much of the north Atlantic population of *Zostera* was decimated in the late 1930s and by the 1960s had recovered [Rasmussen 1978]. One report in England [Ranwell et al. 1974] was that a plugged *Zostera* bed had recovered almost completely after 2–3 years. However, in the following year, the entire bed died. (This might have been due to a host of physical or biological factors.) The recovery rate is sometimes dependent on size of area, since seed stock is often sparse in highly damaged areas and rhizomal growth, especially in *Halodule* and *Syringodium*, is often the chief method of dispersal. Thus a larger area would take longer to recolonize. In general, there appears to be an order of magnitude difference between the restored recovery rate and the natural recovery rate.

Teas [1977] evaluated the mangrove restoration literature. Mangroves are subclimactic intertidal species found throughout the world's subtropical-tropical shorelines which are not completely rocky. There is a great variation between the species of mangrove as to natural recovery; some recover relatively rapidly and others can take as long as many decades (Table I). Recovery rates within 10 years have been seen under restored conditions. Since the mangrove is a large tree, a much longer recovery time is needed than for marsh grasses which do not put as much energy into woody material.

Table I. Summary of Results of Recovery for Restoration Efforts Described in vs Natural

Authors and Date	Location	Number of Species Restored	Type of Species Restored	Place of Restored Species in Community (Climax, Successional, Exotic)
Thorhaug 1976	FL	4	Seagrass	Climax and Successional
Teas 1977	FL and Caribbean	3	Mangrove	Subclimax
Seneca et al. 1976	East U.S. NC coast	1	Marsh grass	Climax
Waring 1976	Pacific NW, U.S.	1	Douglas fir[a]	Successional
Cook 1976	West U.S. desert	6	Shadscale	Climax
Cook 1976	West U.S. foothills	3	Sagebrush, Pines	Successional, Exotic
Cook 1976	West U.S. Inter-elev.	10	Mountain brush	Climax, successional, exotic
Cook 1976	Plains	8	Mixed grass short grass	Climax Climax
Cook 1976	Alpine	3	Grass	Exotic
Brown et al. 1975, 1977, 1978	CO Alpine	7	Alpine grasses and 1 Sedge	Successional, all on low edge of successional scale
Brown et al. 1975, 1977, 1978	Western U.S.	12	Alpine grasses and Sedges	Successional only
Mitchell unpublished	AK arctic	5	Grass	Exotic, successional, climax
Mitchell unpublished	AK boreal	10	Grass	Exotic, successional, climax

[a]In western Oregon and Washington.

"Restoration of Major Plant Communities in the United States" (Thorhaug, 1979)
Recovery Rates

Number of Years to Recovery in Restored Conditions	Number of Years to Partial Recovery in Restoration	Number of Years to Natural Recovery	Success of Plant Recovery in Restoration	Success of Animal Recovery in Restoration Conditions
Thal.–5–6 *Hal.*–3–4 *Syr.*–3–4 *Zos.*–3	3 2 2–5 2	up to 50 10(?) 10(?) 30(?)	80% 100% low	most ? ? ?
?	?	?	?	no data
Spartina–2	1	5 to ? some sites probably never	100%	moderate to good within 5 years
20 to full leaf area	5 to full establishment	20–100	average 75–100%	
30	15	>30	30%	?
7	3	>15 >12	100% 80%	satisfactory ?
7	4	>15	85%	satisfactory
6 8	3 4	>15 >15	100% 100%	satisfactory satisfactory
15	5	>25	30%	?
total recovery not yet attained	5–10	80–150+	50–100%	no data–invertebrate population up
10–25	5–10	100+; maybe250	50–100%	no data–invertebrate population up
Unknown	3 to establish cover. Duration not yet known	5 to >100 w/complete denudation	given sufficient time, >50%	?
Unknown	2 to establish cover. Duration not yet known	50 to >100 to reinstate forest	50–100%	?

Seneca et al. [1976] reviewed marsh grass restoration, especially on the northern Carolina coast, although efforts are now going on from Georgia northward and on the west coast. Marsh grasses are the dominant estuarine intertidal species where rock outcrops do not appear in sheltered coastlines of the temperate United States. The restored areas recover in two years, whereas more than five years is necessary for unassisted recovery, while some areas probably never recover naturally. A 100% success rate of recovery in the work of Seneca et al. is due to extensive experimentation. The associated animal community's recovery is well documented as being excellent within the restored *Spartina* beds. Thus, two to four times more rapid recovery occurs through restoration and one can completely restore areas (Table I).

One of the original techniques for restoring decimated plant communities is reforestation. Waring [1976], as an example of the many worldwide efforts to reestablish forests, summarized the reforestation efforts on the Pacific northwest coast for the Douglas fir. Like the three species of mangroves, this is a large, woody tree, which takes 20–100 years for full natural recovery. Under restoration conditions, partial recovery is achieved in 5 years, while 20 years are necessary for full leaf area to reestablish.

Cook [1976] reviewed a great many types of restoration which he and co-workers have successfully implemented over a long period of time in the western United States, ranging from high alpine to desert plant communities. The number of years to recovery of natural systems ranges from 15 to more than 30. Partial recovery of these same systems takes about 20–30% of the time necessary for natural recovery (Table I). "Fully restored" man-induced recovery requires less than 50% of the time for full "natural" recovery. In general, the grass communities, because of rapid growth and high productivity, are restored in a shorter time than are woody plants. Cook, in his many years of experience, has successfully restored 30 species in various altitudes and climatic conditions. With his efforts, 80–100% success is not unusual. His work is a monument to botanical experimentation with restoration of important species, and should serve as a model to both this and the next generation of restoration scientists.

In general, climax species were the most often used in restoration efforts, although important examples of successional and exotic species appear. Plants with smaller amounts of woody tissue and high productivity, notably grasses, show very rapid rates of restoration recovery. These systems come to complete recovery within 30–50% of the time of the natural recovery rates. Trees and shrubs take longer. High success of recovery has been encountered in many varied species.

The first major issue which was common to all the diverse restoration efforts reported was the problem of understanding the plant communities. A thorough study of the plant species and their ecology occupying the area originally (at minimum, the dominant and subdominant species) should be undertaken. Basic research on important plant communities was then stressed as important long-term background information for any restoration effort. The physiological requirements of the dominant and subdominant species at least in terms of physico-chemical and soil requirements should be determined. The animal community which is being supported by the plant community is important to document. The chemical cycling in the system also may provide basic important information. Ecologists, unlike the "users" of restoration, are aware that documenting such information in many cases takes years of intensive work and often is not available prior to a potential impact. This is unfortunate because many of man's decisions to alter an ecosystem cannot wait until this baseline research is accomplished and must draw on partially completed or in some cases, extremely sketchy information about the functioning of the plant community resource.

The restoration scientist must possess a clear notion of the need and effect of replanting a particular vegetative assemblage before his task can be adequately planned. There are some states such as Oregon [Ternyik in press] which are now requiring an ecological impact evaluation of the restoration itself, because in some cases the restoration of plant species can have considerable effect. The clearer the scientist's understanding of the functioning of the plant community and major plant species within the community and the clearer his notions of the final restoration project, the more fully can he cooperate and participate with decision-makers to tailor the specific restoration site to the "users" and the public's needs in a cost-efficient manner.

It is important to understand the holocoenotical nature of plant community processes [Friederichs 1927; Billings 1964; Webber 1970] as they relate to the restoration and recovery processes. In some cases, major disturbances can result in irreversible damage, due to disruption of critical processes. It is thus important to distinguish reversible and irreversible alterations. An example of irreversibility is thermokarsting (heat-pitting) discussed by Webber and Ives [1978]. As Webber and Ives point out, it is ironic to note that while most processes are equilibrating because they have been selected during evolution, evolution itself is noncyclic as far as individual taxa are concerned and man must thus avoid triggering a disequilibrating process in his impacts. Thermokarsting is triggered by the removal of the insulating and reflective arctic vegetation canopy which, once removed, causes disequilibration of tundra plant communities. The initiation definitely produces devastating effects, which disturbs vegetation underlain by frost-susceptible soils with large amounts of ground ice; this vegetation undergoes an irreversible and severe decline. This is primarily because allogenic processes dominated over the ambogenic processes [Dansereau 1954] and because productivity is low. There was a similar occurrence in north Biscayne Bay, the Atlantic upper limit of extensive *Thalassia* meadows, where 50 years after impact, *Thalassia* revegetation has not occurred, probably due to low incidence fruiting at the upper Atlantic distribution level and man's disturbance of the sediment.

The species planted affect the recovery rate appreciably. Effective and successful efforts have occurred by directly planting climax community or dominant species. There may be a more rapid succession of plant populations in these areas. As a consequence, only several years rather than decades may be required for the climax species to grow back appreciably (seagrass and mangroves are examples). However, in areas where the climax species grow slowly, some of the weedy species can serve an important and immediate physicochemical role by being planted first to prepare the system for the climax species (e.g., tundra vegetation). Either planting or natural processes can introduce the climax species. An example of this is planting more southerly but fast-growing species in Alaska to preserve soil conditions, then letting native species gradually recolonize the area, while the originally planted species dies out. (Caution and considerable prior knowledge must be applied to this technique, for we are all aware of introduced "exotics" which have outcompeted important native species.)

The rate of revegetation is quite an important factor in success of revegetation. Each plant community will have its own growth rate, which appears to vary with:

1. the productivity of the plant itself and the harshness of the environmental conditions (tundra vegetation grows very slowly, whereas temperate salt marshes grow much more rapidly); and
2. the location within the distributional range of the plant species restored (e.g., in the central range of the distribution of deciduous forests or at the limits of deciduous forests in such a manner that in the central portion a more rapid rate of vegetation should occur than at at outer edge).

The length of the growing season at a particular site is important in terms of the rate of revegetation. Clearly arctic plants have a very limited growing season and may take longer to restore than tropical species, many of which grow vigorously all year round.

Successful recovery has been judged not only by the biomass and productivity of the plant community which is restored, but success has been judged on the diversity of the plant community which appears after restoration. Success has also been judged on the basis of diversity and biomass of the animal community which recolonized the restored areas. In some cases, the animal community at certain points in the succession of restoration is even more dense and diverse than the final climax animal community [Seneca et al. 1976]. In other cases, such as tundra restoration, it takes a considerable period of time for the animal community to return and then only selected species return, so that the animal community takes decades to reinhabit [Webber and Ives 1978]. The animal community can return within several years [McLaughlin and Thorhaug 1978; Seneca et al. 1976]. However, animals can also serve as predators and appreciably affect the restoration effort itself. This has been documented in the case of geese feeding on the marsh grass implantations [Garbish, unpublished]. Reforestation efforts where saplings have been eaten by a variety of animals [Waring 1976] are another example of animal interference.

POLITICAL AND LEGAL PROBLEMS

Recovery patterns in restored ecosystems differ from all natural systems discussed before in that political and legal decisions are an intrinsic part of rate of recovery and type of recovery. The species planted, the effort involved as determined by the cost, the area planted and many other factors are decisions determined by the responsible government agency, the user and the restoration scientist. The rate and amount of recovery vary depending on how thoroughly the government agency desires and can permit full restoration, and how well the user follows through with appropriate funds. Unlike natural systems, in which recovery depends heavily on physical conditions and biological potentials, the legal and political aspects are often the dominant factors in restoration recovery. Thus the scientist who attempts to restore must involve himself in this process and bring wisdom and scientific knowledge to bear at strategic points to ensure a rapid recovery.

From a regional viewpoint, mapping of plant communities and planning to reduce damage in highly vulnerable or unique areas are very important as a prerequisite to specific site impact. A master landscape map, such as those now being carried out in the coastal zone, would be highly important to have in areas to be affected so that intended projects can be sited in the best possible area. Thus, siting is an important part of the responsibility of the restoration scientist. A rational assessment of the extent and type of damage from various kinds and degrees of impact and the extent and rate of recovery for the restored community is an important factor to present to political and legal decision-makers. The major thrust of all the restoration work reviewed in the symposium was that careful advance planning is much more desirable from an ecological, economic and political viewpoint. Trade-offs are often necessary. Those projects in which the restoration scientist is asked to come after the project has caused an impact and where no advance planning was done to alleviate the damage are far more costly and less effective than projects where the restoration scientist works hand-in-hand in the initial planning stages along with the decision-makers. Very often, siting areas are chosen primarily on the basis of nonecological criteria; however, better areas for siting which may be slightly more expensive in cost but less expensive in terms of the total "bottom line" of the project, including restoration, may be found.

All those in the symposium who were experienced in plant community restoration emphasized that political and legal realities are integral parts of the process of restoration. Scientists who are not anxious to enter this realm should not involve themselves in restoration, except in very experimental aspects, because there are invariably a great many political and legal problems which must be surmounted before the final restoration can be accomplished. Agreements between a governmental regulating body, the users of the site (often private) and the restoration scientist must be in writing. Terms of what is expected to be planted must be as explicit as possible. A strong legal contract is the best protection the regulating agency and restoration scientist can have against waning enthusiasm for restoration of the "impacting" party once the permit is issued.

CONCLUSIONS

The restoration of plant communities is an important emerging area of ecological research. This is a highly interdisciplinary field with ecological, physiological, agricultural, economic and legal-social components. The scientist must bridge the gap between all of these disciplines in order to be successful.

Plant communities constitute a valuable renewable resource in the United States. Indeed, they served as one of the major factors luring man from the Old World to the "unlimited" resources of North America. Man has intervened drastically by clearing tremendous areas of land for agriculture, urbanization, suburbanization and industry, as well as altering forests and other natural resources (e.g., dammed rivers). Perhaps only eastern China, Western Europe and North Africa compare in extent of alteration of plant communities. Much of this activity has not included rational management of the plant communities themselves, but has been directed toward other goals. Such profound alteration of our plant communities has left us with a wide range of problems.

Recovery processes in restored and natural ecosystems undergo the same physicochemical and biological successional stages. However, in addition to the environmental factors, there are legal, social and economic factors which interact for the final success of the restoration. A few general principles result from examining the foregoing diverse collection of restoration efforts:

1. An understanding of the plant community to be restored is basic to the restoration effort.
2. A clear notion of the need for replanting and the effect of replanting is necessary at the outset in order to make rational decisions which are cost-effective.
3. An understanding of the holocoenotical nature of the plant community is necessary.
4. Choice of the species planted is often very important in the recovery rate and process.
5. The rate of recovery depends on the following ecological factors: (a) fast or slow growing; (b) climax or successional species; (c) climate; (d) where in the range of the species restoration attempt occurs; and (e) productivity of the system.
6. The recolonization rate of animals is important in the successful recovery of the restored ecosystem.
7. The cost is often critical in how much restoration is allowed. However, the efficiency of the restoration process increases (and thus the cost decreases) as repeated restorations are accomplished in various sites of that plant species.
8. The legal and social aspects of the problems often figure significantly in the extent of restoration as does the amount of money required so recovery is highly dependent on societal factors.

It is the hope of the author that more ecologists will become involved in efforts to explore and understand frequently damaged plant communities so that increasingly varied restoration techniques are available for mitigation of man's impact.

REFERENCES

Billings, W. D. *Plants, Man and the Ecosystem* (Belmont, CA: Wadsworth Publishing Co., Inc., 1964), 160 pp.

Brown, R. W., and R. S. Johnston. "Extended field use of screen-covered thermocouple psychrometers," *Agron. J.* 68:995–996 (1976).

Brown, R. W., and R. S. Johnston. "Revegetation of alpine disturbances," *Range Guide* 2(2):2–3 (1976).

Brown, R. W., and R. S. Johnston. "Revegetation of an alpine mine disturbance: Beartooth Plateau, Montana," USDA Forest Service Res. Note INT-207 (1976), 8 pp.

Brown, R. W., and R. S. Johnston. "Rehabilitation of a high elevation mine disturbance," Proc. Workshop on Reveg. of High Altitude Disturbed Lands. Colorado State Univ. Infor. Series No. 28 (1978), pp. 116–130.

Brown, R. W., and R. S. Johnston. "Rehabilitation of disturbed alpine rangelands," First International Rangeland Congress. Soc. Range Manage., Denver, CO (in press).

Brown, R. W., R. S. Johnston and D. A. Johnson. "Rehabilitation of alpine tundra disturbances," *J. Soil Water Conserv.* 33:154–160 (1978).

Brown, R. W., R. S. Johnston, B. Z. Richardson and E. E. Farmer. "Rehabilitation of alpine disturbances: Beartooth Plateau, Montana," Proc. Workshop Reveg. of High Altitude Disturbed Lands. Colorado State Univ. (1976), pp. 58–73.

Brown, R. W., R. S. Johnston and K. Van Cleve. "Rehabilitation problems in alpine and arctic regions," Proc. Symp. Reclamation of Drastically Disturbed Lands (Madison, WI: American Society of Agronomy, 1978), pp. 23–44.

Cairns, J., K. L. Dickson and E. E. Herricks, Eds. *Recovery and Restoration of Damaged Ecosystems* (Charlottesville, VA: University of Virginia Press, 1977).

Cook, C. W. "Surface-mine rehabilitation in the American west," *Environ. Conserv.* 3(3):179–183 (1976).

Dansereau, P. "Studies on central Baffin vegetation. I. Bray Island," *Vegetatio* 5 & 6:329–339 (1954).

Friederichs, K. "Grundsätzliches über die Lebenseinheiten höherer Ordnung und den ökologischen Einheitsfaktor," *Naturwissenschaften* 15:153–157, 182–186 (1927).

Hutnik, R. J., and G. Davis, Eds. *Ecology and Reclamation of Devastated Lands* (New York: Gordon and Breech, 1973), 530 pp.

Johnston, R. S., R. W. Brown and J. Cravens. "Acid mine rehabilitation problems at high elevations," in Am. Soc. Civil Engr. *Symp. Proc. Watershed Management* (Logan, UT: Utah State University Press, 1975), pp. 66–79.

Liverman, J. L. "Some perspectives on land restoration in the United States," *Environ. Conserv.* 4(2):109–114 (1977).

McLaughlin, P., and A. Thorhaug. "Restoration of *Thalassia testudinum*: animal community in a maturing four-year-old site–preliminary results," in *Proc. 5th Annual Conf. on Restoration of Coastal Vegetation in Florida*, D. P. Cole, Ed. (Tampa, FL: Hillsborough Community College Press, 1978), pp. 149–161.

Ranwell, D. S., D. W. Wyer, L. A. Boorman, J. M. Pizzey and R. J. Waters. "*Zostera* transplants in Norfolk and Suffolk, Great Britain," *Aquaculture* 4:185–198 (1974).

Rasmussen, E. "The wasting disease of eelgrass (*Zostera marina*) and its effects on environmental factors and fauna," in *Seagrass Ecosystems*, C. P. McRoy and C. Helfferich, Eds. (New York: Marcel Dekker, Inc., 1977), pp. 1–51.

Schaller, F. W., and P. Sutton, Eds. *Reclamation of Drastically Disturbed Lands* (Madison, WI: American Society of Agronomy, 1979), 768 pp.

Seneca, E. D., S. W. Broome, W. W. Woodhouse, L. M. Cammen and J. T. Lyon. "Establishing *Spartina alterniflora* marsh in North Carolina," *Environ. Conserv.* 3(3):185–188 (1976).

Swank, W., Ed. *Environmental Parameters and Vegetation Establishment on Disturbed Coastal Sites*, U.S. Department of the Interior (in press).

Teas, H. J. "Ecology and restoration of mangrove shorelines in Florida," *Environ. Conserv.* 4(1):51–58 (1977).

Ternyik, W. "Restoration of Pacific Coast marsh grasses," in *Environmental Parameters and Vegetation Establishment on Disturbed Coastal Sites*, W. Swank, Ed., U.S. Department of the Interior (in press).

Thorhaug, A., Ed. *Restoration of Major Plant Communities in the United States* (Amsterdam: Elsevier Press, 1979a).

Thorhaug, A. "The flowering and fruiting of restored *Thalassia* beds: A preliminary note," *Aquatic Bot.* 6:189–192 (1979b).

Thorhaug, A., and C. B. Austin. "Restoration of seagrasses with economic analysis," *Environ. Conserv.* 3(4):259–268 (1976).

Van Kekerix, L. K., R. W. Brown and R. S. Johnston. "The effect of mine spoil treatments on seedling water relations," USDA Forest Service Res. Note INT (1978).

Waring, R. H. "Reforestation in the U.S. Pacific Northwest," *Environ. Conserv.* 3(4):269–272 (1976).

Webber, P. J. "The evangel of the wholeness of the ecosystem," in *Proc. 16th Annual Meeting of the Institute of Environmental Sciences*, M. Lillywhite and C. Martin, Eds., April 1970, No. 17. Boston, MA. 60 pp.

Webber, P. J., and J. D. Ives. "Damage and recovery of tundra vegetation," *Environ. Conserv.* 5(3):171–182 (1978).

CHAPTER 5

INFLUENCE OF ECOSYSTEM STRUCTURE AND PERTURBATION HISTORY ON RECOVERY PROCESSES

Randall M. Peterman*

Institute of Resource Ecology
University of British Columbia
Vancouver, British Columbia
and Canadian Department of Fisheries & Environment

INTRODUCTION

Ecological systems are constantly subjected to a variety of perturbations, both natural and manmade. These disturbances vary in frequency, duration, magnitude and spatial extent. Because biological communities differ in their ability to withstand these perturbations, it is useful to explore how system characteristics affect response capability. This will be done by examining properties of populations and by seeing how the processes which link them together into ecosystems affect stability properties of the populations. The human decision-making structure which imposes changes on the natural community will also be dealt with briefly.

Considerable discussion will be centered around population and ecosystem recovery processes. Recovery here will mean a return to a more "desirable" state from an "undesirable" one. In the context of a management agency's objectives, an undesirable state can be defined in two different ways—for example, too much of some entity during an outbreak of insect pests or diseases, or too few organisms in the case of managed fish or wildlife populations.

This paper is meant to be a general review. Because examples are drawn from different types of ecological systems ranging from agricultural crop

*Present Address: Department of Biological Sciences and Master of Resource Management Program, Simon Fraser University, Burnaby, B.C., Canada V5A 1S6.

diseases through fisheries, general terminology is frequently used to discuss stability and recovery properties. The relatively detailed next section provides definitions and illustrations of these terms which should be sufficient to permit the reader to apply them in the context of any specific ecological system.

It is impossible to provide a survey of all the ways in which ecosystems are perturbed by either natural or manmade agents, or of the numerous system characteristics which affect the ability to recover from disturbance. Therefore only a few properties will be touched upon, particularly those that provide novel interpretations or which support new trends in management.

MULTIPLE EQUILIBRIUM ECOSYSTEMS

Evidence

There is a growing literature which documents existence of ecological systems that have more than one stable equilibrium or attractor. Some studies are based on detailed data from real-world systems [Holling 1973; Sutherland 1974; Southwood and Comins 1976; Clark et al. 1979; McLeod 1979; Peterman et al. 1979]. Others deal with more abstract theoretical cases [Takahashi 1964; Paulik 1973; Austin and Cook 1974; Bazykin 1974; Noy-Meir 1975; May 1977]. The common theme of these studies is that communities which have several nonlinear processes linking system components frequently have multiple equilibria. These systems have the characteristic that each stable equilibrium point for, say, population size, has associated with it a "domain of attraction." Adjacent domains of attraction are separated by a boundary. As long as the system does not cross this boundary, it tends to move back toward the stable equilibrium or attractor of that region. If a boundary is crossed, the system tends toward the stable equilibrium of the new resident attraction domain [see Holling 1973]. In other words, these communities are *not* globally stable; they have distinct limits to the disturbances that can be absorbed without qualitatively changing the complexion of the system. This limited absorptive capacity of multiple equilibrium systems has been called resilience by Holling [1973].

Properties

This multiple equilibrium characteristic has been illustrated in one dimension by examining the shape of the net reproduction curve for a given population [Ricker 1954]. Figure 1 shows the recruitment rate curve for the spruce budworm insect in its host forest system. This curve was derived from a model which combined effects of field measured processes such as parasitism, bird predation and larval insect competition [Morris 1963; Clark et al. 1979]. A very similar curve has been derived to describe some fish populations [Paulik 1973; Peterman 1977]. Wherever such a recruitment curve crosses the R = 1 line, there is a potential equilibrium. In this case points A and C are stable and "attractant" and B is unstable or "repellent". Population densities

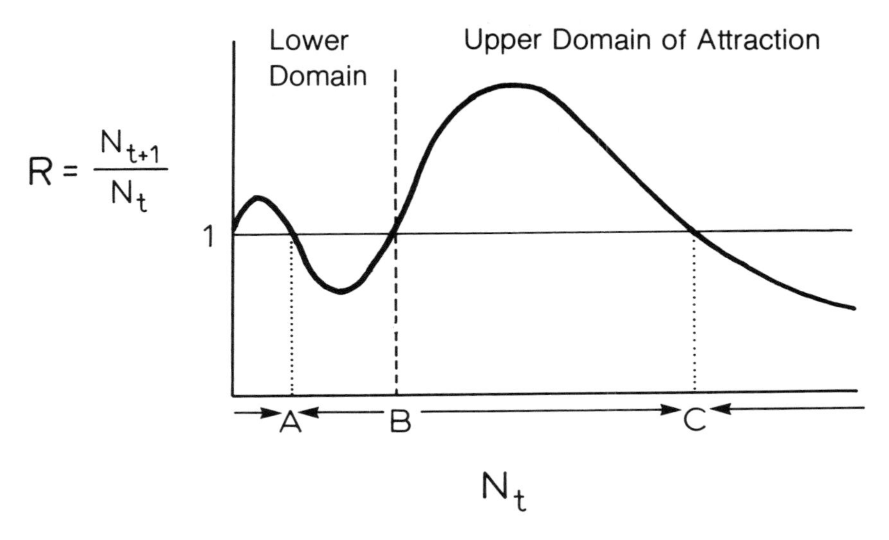

Figure 1. A multiple equilibrium recruitment curve similar in form to that derived for the spruce budworm-forest system and for Pacific salmon. When population growth rate (R) as shown by the curve is greater than 1, density (N_t) increases; if $R < 1$, density decreases. A and C are stable equilibria or attractors and the population density B represents the boundary separating the upper and lower domains of attraction.

above B tend to move toward C while those below B tend to move toward A. Thus, the density at B forms the boundary between two separate domains of attraction. Populations will likely never achieve a constant density at one of the stable equilibria, but the locations of the attractors and repellor(s) dictate the direction and rate of change in the population. They thereby create a "stability landscape" of valleys and ridges through which the population travels.

There can be four main types of perturbations of populations:

1. perturbations in variables such as population density;
2. changes in parameters such as the slope of the fecundity vs. density relation;
3. changes in exogenous variables such as temperature; and
4. "structural" changes in the community through addition or loss of species.

Each type of perturbation has its own particular effect on multiple equilibrium systems, as illustrated below.

The type of recruitment curve in Figure 1 exhibits five properties that are relevant to recovery of disturbed ecosystems. First, for a perturbation of a variable such as population density, return to the former resident domain does not automatically occur along with removal of the perturbing force that shifted the system to a new region. If the recruitment curve of Figure 1 represents a fish population, then as long as harvesting of a population resident in the upper domain of attraction does not push the population across the boundary, it will remain in the more desirable domain of attraction and recovery from that perturbation will be normal. However, if overharvesting occurs and the system crosses the boundary into the lower domain, then recovery will *not* occur by simply stopping exploitation. The density will have to be actively moved back above point B.

This property of boundary crossing can create a time series plot like Figure 2, which is the pattern of data on salmon [Peterman 1977], clupeoid fisheries [Murphy 1977] and fish of the Great Lakes [Christie 1974; Loftus 1976]. In the third case, after exploitation was greatly reduced, the previously large stocks that were overharvested stayed small, in part due to the combined effects of new predators and competition from other fish in the community. The work of Niering and Goodwin [1974] also depicts a likely case of multiple equilibria in a plant community, which shows the same pattern through time. Trees growing in a power line right-of-way were sprayed with herbicide for a few years in succession, and in the 15 years following the cessation of spraying the community has been dominated by shrubs which prevent establishment of tree seedlings. In this situation, the lower domain is the desirable state from the standpoint of management goals. Many other examples exist for this type of situation, e.g., diseases, insect pests [May 1977].

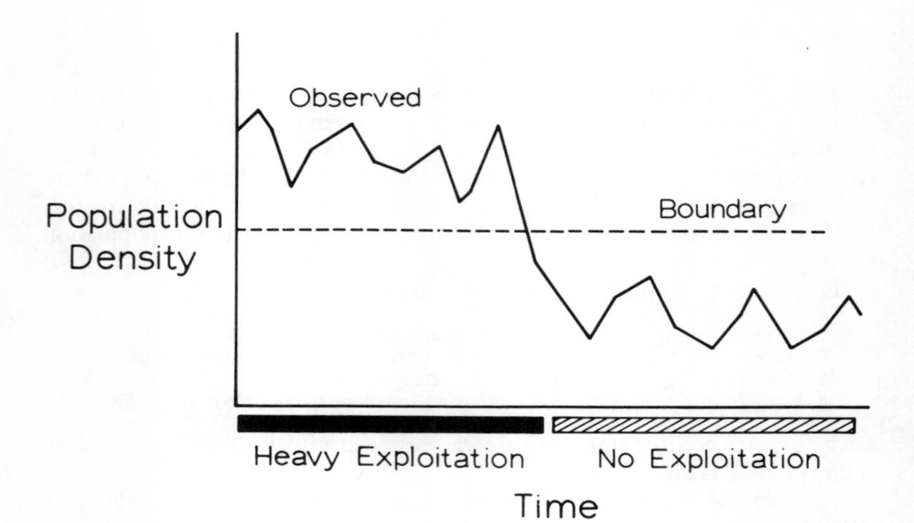

Figure 2. Temporal behavior of a hypothetical exploited population which has a reproduction curve as in Figure 1 and which starts out in upper domain. Solid line is observed population density, dashed line is density of the boundary population which separates the two domains of attraction.

The second property of these systems is that perturbations in parameters or driving variables can move the critical boundary by modifying the shape of the recruitment curve (Figure 3). Examples which have been explored in some of the modeling papers cited above are poor weather conditions, changing food quality or intensification of mortality on fishes by invasion of predators. In each case, the boundary movement can mean that a population density which used to be well within a desirable domain may suddenly be found in an undesirable domain (Figure 4). Other examples of moving boundaries are given by Peterman et al. [1979].

Third, management regimes can in themselves cause a change in the stability landscape by altering parameters or variables. A simple demonstration of this is given by comparing the age-specific mortality rates in natural and managed systems. Manipulation of populations through harvesting often (but not always) removes the older and what are usually the most fecund individuals. This can greatly affect the population's ability to survive future perturbations. In the specific case of fisheries, exploitation often results in changes in age structure and age at first reproduction [Beverton and Holt 1957; Sergeant 1966; Lockyer 1977]. There may even be an evolutionary response in growth patterns to fishing pressures [Handford et al. 1977; Ricker et al. 1978]. Because age structure, age at first reproduction and growth can greatly influence reproductive output, management disturbances can have enormous influence on the ability of a population to absorb any perturbation. Failure to take into account such indirect system changes may reduce the probability that a system will recover from a disturbance after any given remedial action.

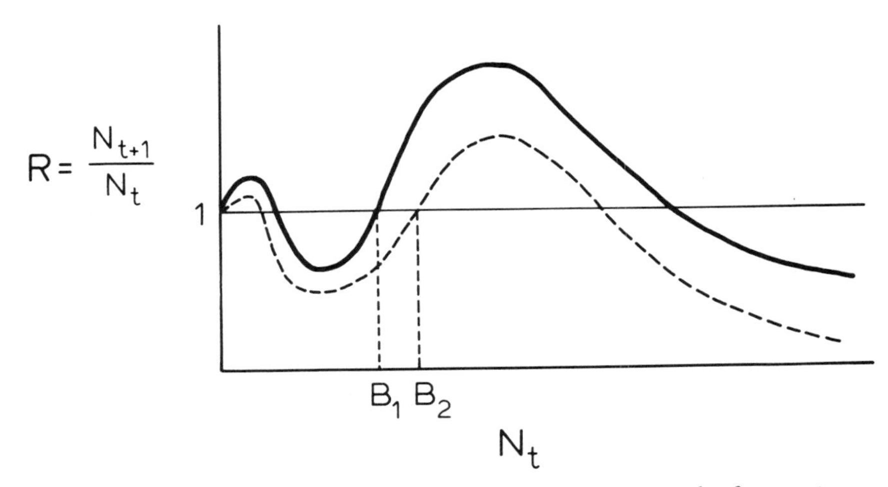

Figure 3. Boundary population density B_1 increases to B_2 as a result of parameter or driving variable changes. Other modifications in these two sources of change can cause a decrease in the boundary population density.

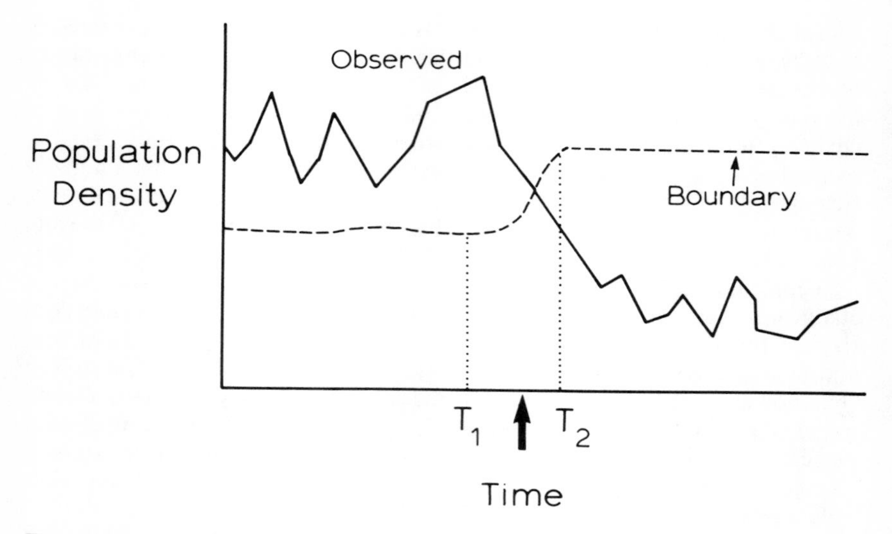

Figure 4. A hypothetical time series of an unexploited population in which perturbations cause a movement in the boundary population size as in Figure 3. T_1 and T_2 are the times at which the boundaries are at B_1 and B_2 in Figure 3, respectively. Solid line is observed population density and dashed line is the boundary density.

Fourth, approach to the boundary of a stability region may be undetectable to the observer. Multiple equilibrium systems can cope with successive increments in harvest rates or other changes without any visible effects, if the wrong variables are being monitored. But one final increment moves the system across a boundary and there is a rapid change in system character. For example, in the spruce budworm-forest community, annual changes in forest maturity have virtually no detectable effect on endemic insect populations, but once a critical foliage density is surpassed, an outbreak follows [Clark et al. 1979].

Finally, management regimes which manipulate populations to achieve some objective often keep an ecological system in a new part of the resident domain of attraction. This move can thereby change the likelihood that a boundary will be crossed. The setting of environmental health standards for various pollutants is designed to do just this, to avoid a supposedly critical boundary which separates desirable and undesirable states [Fiering and Holling 1974]. A counterproductive example is provided by standard management practice in salmon fisheries, which actually increases the likelihood of crossing a boundary by attempting to keep the system in the most productive state. A major goal in salmon management since the mid-1960s has been to let a fixed number of fish escape the fishing nets, this number being the one that would result in the maximum sustainable yield (MSY). However, it has been shown [Ricker 1963; Peterman 1977] that fish populations which are poised on the MSY point are more likely to suffer catastrophic declines in production than are stocks exploited less intensely, because the system is being held nearer the boundary of the lower domain of attraction.

The foregoing points raise several important issues. Attempts to induce recovery of disturbed ecosystems may be ineffective if a multiple equilibrium situation exists but is not taken into account. Next, the form and magnitude of the processes (competition, predation, reproduction, etc.) that link all the system components determine the shape of the stability landscape and thereby dictate the type of response that will result from a given disturbance. Man-caused disturbances impose an additional source of change to these life history parameters and the stability landscape can be modified in such a way that man's recovery attempts are undermined. Finally, natural and manmade perturbations can have the same qualitative effect on system structure and behavior in terms of changes in location of the system in the stability landscape. However, manmade disturbances may be of such frequency, magnitude, duration and spatial extent that novel responses, such as evolutionary ones can result.

PERTURBATION HISTORY AND RESILIENCE

Mechanisms

As already discussed, ecosystems vary in their ability to survive a given perturbation with little change, in part because of differences in system structure. These distinctions are reflected in the shape and size of domains of attraction of the component populations, and these domains are in turn products of the parameters of component system processes, as schematically illustrated in Figure 5. Years of selective breeding experience with agricultural plants and animals have demonstrated that certain life history characteristics can be rapidly changed by intensive selection. Thus there is a series of connections between natural selection, life history parameters, stability characteristics and population dynamics. And, as discussed above, management actions can modify stability properties by entering this loop at various points.

Because of this cause-effect relation between system components and stability characteristics, it is important to determine if perturbations can change a community's likelihood of surviving future disturbances. This appears to be possible because there is one final causal link that closes the loop in Figure 5, the link between population dynamics and selection pressures. A few studies of coevolution demonstrate the existence of this link. The Australian myxomatosis-rabbit complex is a classic example. Rabbit abundance, rate of infection and virulence of the pathogen all changed drastically as the virus and its host went through a series of evolutionary adjustments [Fenner and Ratcliffe 1965]. In similar experimental studies with the housefly and a parasite, Pimentel claimed that after a number of generations the host became more resistant to the parasite, the parasite less

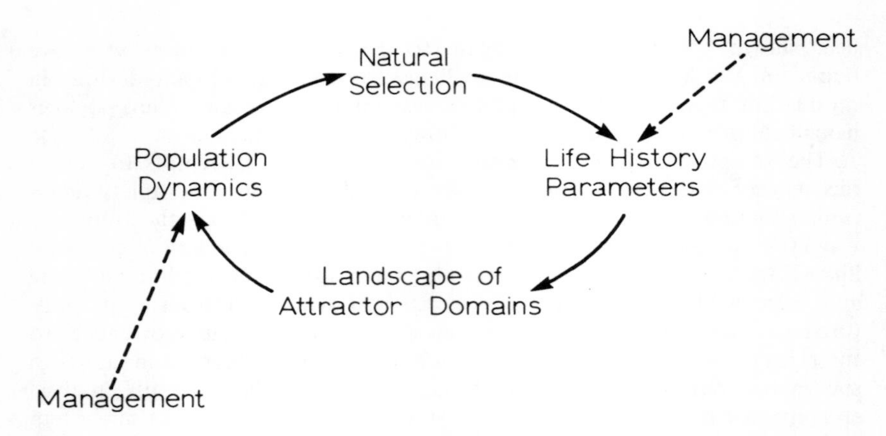

Figure 5. Natural selection acts on life history parameters, which, acting together, dictate the number, size and shape of domains of attraction. These domains in turn affect population dynamics and responses to perturbations by drawing the population towards appropriate equilibria. Population changes can change intensity or direction of natural selection, which can ultimately alter the stability landscape. Management actions also impinge on the system at various points.

virulent, and the population fluctuations less violent [Pimentel et al. 1963; Pimentel and Stone 1968]. Finally, the parasite *Mesoleius tenthredinis* was introduced into Canada to control the larch sawfly and was initially reported to be highly successful in reducing outbreak populations [Muldrew 1953]. However, within 20–30 years it was found that the host sawfly had developed the ability to encapsulate the parasite eggs [Muldrew 1953]and the host population again reverted to preparasite levels [Turnock 1972].

Each of these examples illustrates the connection between population dynamics and selection pressures. When host population densities were drastically reduced, less virulent pathogens were selected. This resulted in modification of the host's stability properties and population fluctuations.

Examples

Theoretically, then, the historical pattern of disturbances can change a system's resilience. Several natural and man-manipulated communities exemplify a more general link between perturbation history and ability to absorb future stresses. First is the forest-fuel-fire complex commonly found in western North America. Frequent lightning-started fires used to keep underbrush or accumulated deadwood at low levels but man's fire control practices have drastically reduced the frequency and extent of this periodic removal of fuel material. This has led to the dangerous situation where the occasional fire that does get out of control causes a catastrophic conflagration which kills the overstory of trees. There appears to be an inverse relation between fire frequency and fire intensity [Vogl 1977]. The value of frequent fires in maintaining the forest's ability to survive future fires has now become widely recognized and management policies have been set which start controlled burns and which permit some natural fires to burn unhindered [Wright and Heinselman 1973; Kozlowski and Ahlgren 1974]. Thus, the role of past perturbations in maintaining the forest's resilience has been clearly recognized.

Watt [1968] showed in his analysis of Canadian Forest Insect and Disease Survey data that insects which inhabit regions with less variable climates are *more* sensitive to unit changes in weather than are insects living in more variable climates. Watt concluded that because of their history of greater climatic variability, insects in the latter regions have evolved a lower sensitivity to variations in weather and hence are more able to persist after extreme conditions are encountered.

Two more familiar examples from human systems also illustrate this phenomenon. First is the disease-immune complex. With some diseases, a single exposure will create lifelong immunity. With others, periodic exposure through vaccinations or personal contact may be required to maintain a person's ability to survive future exposure to that disease. In such cases, the less frequent the exposure to disease, the higher the probability that the exposure will be fatal.

Many human institutions behave in a similar way. Watt [1974] describes the "Titanic effect," in which the *less* likely a disaster is to occur, the more disastrous will be its consequences when it does occur because people will not be well prepared. In this sense, periodic perturbations (floods, energy shortages, jumps in fertilizer costs, etc.) can be beneficial because they force institutions to evolve mechanisms to handle future disturbances of those types.

REDEFINING THE "OPTIMAL" MANAGEMENT REGIME

It is likely that only some ecological systems have the characteristics outlined above. In these cases, the way we manage ecosystems and impose our own perturbations could influence the ability of these systems to persist in the face of future disturbances, either natural or human. Attempts to hold dynamic systems in static states which are ideal from a management point of view could lead to a decrease in resilience and greater risk of disaster. Similar effects could result from a regular pattern of perturbations.

These observations lead to the suggestion that traditionally defined "optimal" management regimes should be redefined to take into account risks arising from system resilience properties. An explicit value should be attached to the resilience capabilities of the system [to define measurements of these abilities see Fiering and Holling 1974; Beddington et al. 1976; Peterman 1977]. Then, resource managers can trade off traditional benefits of manipulating a system with its sensitivity to perturbations and associated risks.

Examples

The fire management situation described above is perhaps the best known case of explicit recognition of the "resilience value" of frequent perturbations. Forest managers would prefer effective fire control practices that permit simple achievement of their particular management goals. But this results in a high risk situation whereby highly improbable fires that do get out of control can create extremely high costs. Now some foresters realize that higher frequency of fires ultimately means lower cost of fires. Using Holling and Clark's [1975] terminology, these foresters have traded a fail-safe approach for a safe-fail one, i.e., one that is less costly in the inevitable event that fire control measures fail.

A similar approach has been proposed for management of Pacific salmon fisheries. As noted earlier, most of these stocks are managed with the goal of obtaining MSY, but doing this puts the system in a precarious position that is particularly vulnerable to slight perturbations or miscalculations. Because overexploited salmon stocks are not easily rehabilitated [Larkin 1974], a strong case can be made for reduction in harvest rates below the optimal MSY rate in order to give the population more leeway in the face of unknown future disturbances [Peterman 1977].

The two examples above show how system resilience can be increased by *managing* to meet some redefined objective which accounts for risks. A similar result can be obtained by *designing* a system to operate with this new goal in mind. Genotypic manipulation of agricultural crops through selective breeding programs is a prime example. Cereal crop varieties that were resistant to specific rusts were very successful and productive for 5–20 years, but they all failed when new rust strains evolved [Johnson 1953]. These cereal varieties were initially so successful that they quickly became used on a continent-wide basis at the expense of numerous "wild genotypes." This resulted in the precarious situation that crop production was optimal in the absence of pathogens, but when the rapid recombination mechanism of rusts resulted in new rust races, disastrous outbreaks of disease occurred [Day 1977; National Academy of Science 1972]. By concentrating on maximizing yield and resistance of plants, breeders were unwittingly selecting for low adaptability or resilience [Browning and Frey 1969]. In fact, the sequential introduction of new resistant wheat varieties merely speeded up evolution of more pathogenic rusts [Johnson 1961].

Various solutions to this problem are presently being tried and they all have the common feature that the crops are manipulated *away from* the short-term, economically optimal growing pattern. Crops can be a mixture of different resistant genotypes [Leonard 1969], they can be planted in a heterogeneous spatial array [Van der Plank 1968; Browning and Frey 1969] or they can be alternated in time. All of these methods result in some additional cost to the grower but with some recognized benefit in terms of lowered chance of catastrophic disease outbreak. Attempts to breed crops resistant to numerous different rusts lead to less productive and competitive plants [Schwarzbach 1975; Van der Plank 1975].

Institutional Implications

The area of selective crop breeding provides us with another lesson relevant to the topic of system resilience and recovery. This deals with how to get management institutions to recognize the need to maintain system resilience. According to Marshall [1977] there has been a large volume of literature since the 1930s on the dangers of decreasing genetic variability of plant crops, and yet the trend toward more crop uniformity continued unabated. People began to take the issue seriously only after a series of costly and widespread plant disease outbreaks. The same was true of the fire management situation described above. The large number of destructive fires finally forced a recognition of the beneficial role of some small frequent fires. The lesson appears to be that only after we have manipulated resource systems to become more efficient in terms of our narrowly defined short term goals do we begin to learn that systems become more susceptible to inevitable perturbations.

This brings us back to Watt's "Titanic effect." Disturbances can be "creative" in that systems (natural plus management structure) will respond by evolving mechanisms to cope with future disturbances of that type. The costs of these disturbances may be worth incurring when they are compared with the higher costs that could result from perturbing a previously undisturbed system.

Adaptive Management

One final argument can be made by analogy to justify disturbing a system away from the traditionally defined "optimum" condition. This is done by attaching a value to information gained in a perturbation experiment, just as we attach a value to increased resilience. The idea behind adaptive management is to perturb the system away from what is the best guess of the optimum, given the management objective, and to gain information about the true optimum by monitoring the response. Foresters, uncertain about the best planting densities and thinning regimes, have run plots under different conditions of these two variables [e.g., Cromer and Pawsey 1957]. After several dozen years it has become clear which scheme best achieves a particular management goal, and this has not always matched what the theoreticians predicted [Marsh 1957; Smith 1958]. This brute force empirical approach thus yields valuable information that would not have been produced had the theoretically best management practices been followed.

Similarly, some investigators have proposed that fish and whale populations which have only been observed over a narrow range of conditions be managed by manipulating them in different ways and observing results to learn more about the underlying population processes [Larkin 1972; Walters and Hilborn 1976; Holt 1977; Peterman 1977]. In all these cases the value of the information gained is explicitly taken into account when deciding whether to perturb the system away from the presumed optimum [Peterman et al. 1978; Silvert 1978]. Similar arguments for adaptive management have been put forth in the context of environmental assessment [Holling et al. 1978].

An exact parallel can now be drawn with our original situation of disturbing systems to maintain their resilience. As suggested earlier, we should be willing to decrease the short-term productivity of our managed systems to increase the likelihood of their persistence. The tradeoff that is being made between present harvests and experimental information should be made just as explicitly as the trade-off between short-term management goals and resilience.

CONCLUSIONS

Manmade disturbances can cause an evolutionary change in an ecosystem's stability landscape through selection on parameters of various population processes. Through this and other mechanisms, a history of perturbations can increase the probability that the system will survive future stresses of that type. But human management often removes these disturbances and tries to keep the system near its peak productivity. In many cases, this is done at the expense of the system's resilience, or ability to persist after disturbance. The only way to avoid falling into the trap of managing to maximize short-term productivity is to clearly recognize the value of resilience capability.

ACKNOWLEDGMENTS

Many thanks are due to W. C. Clark, C. S. Holling and R. Fleming who, through many discussions, helped clarify several ideas presented here. Useful comments on a draft manuscript were provided by J. Anderson, A. Auclair, A. R. E. Sinclair and J. M. McLeod.

REFERENCES

Austin, M. P., and B. G. Cook. "Ecosystem stability: a result from an abstract simulation," *J. Theor. Biol.* 45:435–458 (1974).

Bazykin, A. D. "Volterra's system and the Michaelis-Menten equation" (in Russian), in *Problems in Mathematical Genetics*, V. A. Ratner, Ed. (Novosibirsk: USSR Academy of Science, 1974), pp. 103–143.

Beddington, J. R., C. A. Free and J. H. Lawton. "Concepts of stability and resilience in predator-prey models," *J. Anim. Ecol.* 45(3):791–816 (1976).

Beverton, R. J. H., and S. J. Holt. "On the dynamics of exploited fish populations," *Fish. Invest. Minist. Agric. Fish. Food (Gr. Brit.) Ser. II*, Vol. 19, 533 pp. (1957).

Browning, J. A., and K. H. Frey. "Multiline cultivars as a means of disease control," *Ann. Rev. Phytopathol.* 7:355–382 (1969).

Christie, W. J. "Changes in fish species composition of the Great Lakes," *J. Fish. Res. Board Can.* 31:827–854 (1974).

Clark, W. C., D. D. Jones and C. S. Holling. "Lessons for ecological policy design: A case study of ecosystem management," *Ecol. Modelling* 7:1–53 (1979).

Cromer, D. A. N., and C. K. Pawsey. "Initial spacing and growth of *Pinus radiata*," Bull. No. 36, Forestry and Timber Bureau, Canberra, Australia (1957), 42 pp.

Day, P. R., Ed. "The genetic basis of epidemics in agriculture," *Ann. N.Y. Acad. Soc.*, Vol. 287, 440 pp. (1977).

Fenner, F., and F. N. Ratcliffe. *Myxomatosis* (New York: Cambridge University Press, 1965), 379 pp.

Fiering, M. B., and C. S. Holling. "Management and standards for perturbed ecosystems," *Agro-Ecosystems* 1:301–321 (1974).

Handford, P., G. Bell and T. Reimchen. "A gillnet fishery considered as an experiment in artificial selection," *J. Fish. Res. Board Can.* 34:954–961 (1977).

Holling, C. S. "Resilience and stability of ecological systems," *Ann. Rev. Ecol. Syst.* 4:1–23 (1973).

Holling, C. S., and W. C. Clark. "Notes toward a science of ecological management," in *Unifying Concepts in Ecology*, W. H. van Dobben and R. H. Lowe-McConnell, Eds. (The Hague: Dr. W. Junk B. V. Publishers, 1975), pp. 247–251.

Holling, C. S., A. Bazykin, P. Bunnell, W. C. Clark, G. C. Gallopin, J. Gross, R. Hilborn, D. D. Jones, R. M. Peterman, J. E. Rabinovich, J. H. Steele and C. J. Walters. *Adaptive Environmental Assessment and Management* (Chichester, England: John Wiley and Sons, 1978), 377 pp.

Holt, S. J. "Whale management policy," 27th Report of International Whaling Commission. Cambridge, pp. 133–137 (1977).

Johnson, T. "Variation in the rusts of cereals," *Biol. Rev.* 28:105–157 (1953).

Johnson, T. "Man-guided evolution in plant rusts," *Science* 133(3450): 357–362 (1961).

Kozlowski, T. G., and C. E. Ahlgren, Eds. *Fire and Ecosystems* (New York: Academic Press, Inc., 1974), 542 pp.

Larkin, P. A. "A confidential memorandum on fisheries science," in *World Fisheries Policy: Multidisciplinary Views*, B. J. Rothschild, Ed. (Seattle, WA: University of Washington Press, 1972), pp. 189–197.

Larkin, P. A. "Play it again Sam—an essay on salmon enhancement," *J. Fish. Res. Board Can.* 31:1433–1459 (1974).

Leonard, K. J. "Selection in heterogeneous populations of *Puccinia graminis* f. sp. avenae," *Phytopathology* 59:1851–1857 (1969).

Lockyer, C. "A preliminary study of variations in age at sexual maturity of the fin whale with year class, in six areas of the southern hemisphere," 27th Report of International Whaling Commission, pp. 141–147 (1977).

Loftus, K. H. "Science for Canada's fisheries rehabilitation needs," *J. Fish. Res. Board Can.* 33:1822–1857 (1976).

Marsh, E. K. "Some preliminary results from O'Connor correlated curve trend (C.C.T.) experiments on thinnings and espacements and their practical significance," 7th British Commonwealth Forestry Conference, Govt. Printer, Pretoria, South Africa (1957), 21 pp.

Marshall, D. R. "The advantages and hazards of genetic homogeneity," in *The Genetic Basis of Epidemics in Agriculture*, P. R. Day, Ed., *Ann. N.Y. Acad. Sci.* 287:1–20 (1977).

May, R. M. "Thresholds and breakpoints in ecosystems with a multiplicity of stable states," *Nature* 269(5628):471–477 (1977).

McLeod, J. M. "Discontinuous stability in a sawfly life system and its relevance to pest management strategies," in *Current Topics in Forest Entomology*, W. E. Waters, Ed., Selected Papers from XV Intern. Cong. Ent., Washington, DC, U.S. Forest Service, General Tech. Report WO-8 (1979), pp. 68–81.

Morris, R. F., Ed. "The dynamics of epidemic spruce budworm populations," *Mem. Entom. Soc. Canada*, No. 31 (1963).

Muldrew, J. A. "The natural immunity of the larch sawfly (*Pristiphora erichsonii*) to the introduced parasite *Mesoleius tenthredinis*, in Manitoba and Saskatchewan," *Can. J. Zool.* 31(4):313–332 (1953).

Murphy, C. I. "Clupeoids," in *Fish Population Dynamics*, J. A. Gulland, Ed. (London: John Wiley and Sons, 1977), pp. 283–308.

National Academy of Science. *Genetic Vulnerability of Major Crops* (Washington, DC: National Academy of Science, 1972).

Niering, W. A., and R. H. Goodwin. "Creation of relatively stable shrublands with herbicides: arresting "succession" on rights-of-way and pastureland," *Ecology* 55:784–795 (1974).

Noy-Meir, I. "Stability of grazing systems: an application of predator-prey graphs," *J. Ecol.* 63:459–481 (1975).

Paulik, G. "Studies of the possible form of the stock and recruitment curve," in *Fish Stocks and Recruitment*, B. B. Parrish, Ed., *Rapp. Proc.-V. Reun. Cons. Int. Explor. Mer.* 164:302–315 (1973).

Peterman, R. M. "A simple mechanism that causes collapsing stability regions in exploited salmonid populations," *J. Fish. Res. Board Can.* 34(8):1130–1142 (1977).

Peterman, R. M., W. C. Clark and C. S. Holling. "The dynamics of resilience: shifting stability domains in fish and insect systems," in *Population Dynamics*, R. M. Anderson, R. E. Taylor, B. D. Turner, Eds. (Oxford: Blackwell Scientific Publishers, in press).

Peterman, R. M., C. J. Walters and R. Hilborn. "Systems analysis of Pacific salmon management problems," in *Adaptive Environmental Assessment and Management*, C. S. Holling, Ed. (Chichester, England: John Wiley and Sons, 1978), pp. 183–214.

Pimentel, D., and F. S. Stone. "Evolution and population ecology of parasite-host systems," *Can. Entom.* 100:655–662 (1968).

Pimentel, D., W. P. Nagel and J. L. Madden. "Space-time structure of the environment and the survival of parasite-host systems," *Am. Nat.* 97:141–167 (1963).

Ricker, W. E. "Stock and recruitment," *J. Fish. Res. Board Can.* 1:559–623 (1954).

Ricker, W. E. "Big effects from small causes: two examples from fish population dynamics," *J. Fish. Res. Board Can.* 20:257–264 (1963).

Ricker, W. E., H. T. Bilton and K. V. Aro. "Causes of the decrease in size of pink slamon (*Oncorhynchus gorbuscha*)," Fisheries & Marine Serv. Tech. Rep. No. 820 (1978), 93 pp.

Schwarzbach, E. "The pleiotrophic effects of the ml-o gene and their implications in breeding," in *Barley Genetics III*, Proc. 3rd Int. Barley Genetics Symp., Garching (1975), pp. 440–445.

Sergeant, D. E. "Reproductive rates of harp seals, *Pagophilus groenlandicus* (Erxleben)," *J. Fish. Res. Board Can.* 23:757–766 (1966).

Silvert, W. "The price of knowledge: fisheries management as a research tool," *J. Fish. Res. Board Can.* 35:208–212 (1978).

Smith, J. H. G. "Better yields through wider spacing," *J. Forestry* 56(7): 492–497 (1958).

Southwood, T. R. E., and H. N. Comins. "A synoptic population model," *J. Anim. Ecol.* 45:949–965 (1976).

Sutherland, J. P. "Multiple stable points in natural communities," *Am. Nat.* 108(964):859–873 (1974).

Takahashi, F. "Reproduction curve with two equilibrium points: a consideration on the fluctuation of insect population," *Res. Population Ecol.* 6:28–36 (1964).

Turnock, W. J. "Geographical and historical variability in population patterns and life systems of the larch sawfly (Hymenoptera: Tenthredinidae)," *Can. Entomol.* 104:1883–1900 (1972).

Van der Plank, J. E. *Disease Resistance in Plants* (New York: Academic Press, Inc., 1968), 206 pp.

Van der Plank, J. E. *Principles of Plant Infection* (New York: Academic Press, Inc., 1975).

Vogl, R. J. "Fire: a destructive menace or natural process?" in *Recovery and Restoration of Damaged Ecosystems*, J. Cairns, Ed. (Charlottesville, VA: University Press of Virginia, 1977), pp. 261–289.

Walters, C. J., and R. Hilborn. "Adaptive control of fishing systems," *J. Fish. Res. Board Can.* 33:145–159 (1976).

Watt, K. E. F. "A computer approach to analysis of data on weather, population fluctuations, and disease," in *Biometeorology*, W. P. Lowry, Ed. (Corvallis, OR: Oregon State University Press, 1968), pp. 145–159.

Watt, K. E. F. *The Titanic Effect* (Stamford, CT: Sinauer and Associates, 1974), 268 pp.

Wright, H. E., and M. L. Heinselman. "The ecological role of fire in natural conifer forests of western and northern America," *Quat. Res.* 3(3):317–513 (1973).

CHAPTER 6

MULTIVARIATE QUANTIFICATION OF COMMUNITY RECOVERY

Stephen A. Bloom

Department of Zoology
University of Florida
Gainesville, Florida 32611

INTRODUCTION

As the ecological impact of human activities on natural communities has increased, so has the interest in quantifying and understanding changes and patterns of changes within those communities. Given a distinct and temporally limited perturbation, there are three major questions:

1. Does the community recover or change states (*sensu* Sutherland 1974)?
2. If the community recovers, how rapid and how complete is the recovery?
3. What is the path of the recovery after the perturbation?

The form of these questions implies a reasonable knowledge of the community before and after perturbation. The existence of such data will be an underlying assumption of the technique presented here.

Any analytical technique, if it is to be of any real utility, must possess certain features. Chief among these are the maximum use of available information and the ability to be easily grasped. If the user of a model or a technique does not have an intuitive feel for the assumptions, advantages and limitations of the tool that he is using, the potential for misuse is high. While the technique presented here uses relatively sophisticated mathematical procedures, these procedures are readily available as computer packages [Bloom et al. 1977] and are intuitively obvious.

The usual types of quantitative biotic information gathered about natural communities are the number of species, the species composition (the taxa represented), and the number of individuals per species in some defined sample size. There are several techniques addressed to recovery problems which use one or more of these informational types. For example, island biogeography [MacArthur and Wilson 1967; Simberloff 1974] deals with species composition and numbers, while multivariate approaches [Allen et al. 1977] deal with all three informational types.

The technique presented here is an extension of the multivariate approach combined with a geometrical and hypothesis-testing statistical analysis. The basic technique is simple and conforms closely to general ecological concepts of community stability and resilience [Holling 1973]. The similarity (dissimilarity) of all pre- and postperturbation samples are represented as intersample distances in a three-dimensional ordination system. The preperturbation samples are defined as a cluster and the statistical envelope which encloses the cluster is calculated. Sequentially, each postperturbation sample is tested to see whether it lies within the envelope (statistically recovered) and how far it lies from the nearest edge of the envelope (the distance to recovery). The distance to recovery is then plotted through time to reveal patterns of multistable states, successional stages, speed to recovery and the path of recovery in ordination space.

DESCRIPTION OF DATA SET

The development of the multivariate technique requires the use of a data set featuring normal variation, a reasonable time frame, a distinct perturbation and before and after quantitative monitoring of the community. Such a data set appeared to exist in two studies of a sandy intertidal habitat at Courtney Campbell Causeway in Old Tampa Bay, Florida. The first study [Bloom et al. 1972] represented a 9-month survey of the site beginning in September 1968. The second was an intensive 24-month monitoring of the same site after virtual defaunation due to Red Tide in August 1971 [Dauer and Simon 1976]. Because of differences in sampling techniques, sieve sizes and taxonomic expertise, the data from the two studies were not comparable and the studies do not represent a legitimate before-and-after perturbation data set. However, to develop the technique, the second year of data from the Dauer and Simon [1976] study was considered to be "preperturbation". The results presented here are thus not a legitimate test of the technique but are simply an example of how the analysis works. Given the definition of preperturbation, the analysis should show a distinct trend towards "recovery" if the technique is valid.

DESCRIPTION OF ANALYTICAL PROCEDURE

Ordination

The initial data matrix (species by sample with counts per square meter filling the matrix) is reduced and converted to an n-dimensional representation of intersample distances by a principal coordinate analysis (PCOR) [Gower 1966; Sneath and Sokal 1973]. PCOR requires a unit-variance standardization [Whittaker 1973] but does not specify any particular data transformation. The output of a PCOR includes a table of samples by principal axes. The axes are ranked in descending order by the amount of variance in the original data "explained" by the axes. At the present stage of development, only the first three axes are considered. By plotting the values of the various samples on the three axes, a three-dimensional representation of the intersample distances is generated.

In the data set used here, the 153 species by 24 samples (monthly samples) was so treated initially using a square-root transformation. The results on principal axes I and II are presented in Figure 1. Samples 1 and 12 are defined as preperturbation but are actually the second year of monitoring. Sample 13 represents the community state immediately after perturbation and each successive sample represents the community state in ordination space one month later.

Cluster Definition

The preperturbation samples are defined as a cluster. By analyzing the positions of these preperturbation samples, a 95% rejection envelope can be calculated. The envelope defines all points in ordination space which are statistically indistinguishable ($\alpha = 0.05$) from the preperturbation samples. Thus if a postperturbation sample should fall into the envelope, it has statistically recovered, i.e., can no longer be said to be statistically distinct from the preperturbation community state. Obviously, the cohesiveness of the preperturbation cluster will directly affect the probability of any point being found to be statistically distinct from the cluster. If the variance is low between preperturbation samples, the envelope will enclose a relatively small volume of space and a randomly distributed point will most likely fall outside of the cluster. On the other hand, if the preperturbation variance is high, a large volume will be enclosed and virtually any community state will be indistinguishable from the preperturbation community state. The value of replicate sampling in realistically defining the preperturbation state is obvious.

The amount of information and, hence, the importance of any axis is theoretically related to the amount of variance "explained" by the axis. The coordinates of samples on an axis can be weighed by the amount of variance "explained" by that axis by multiplying the coordinate by an axis weighing constant. The axis weighing constant is simply the amount of variance accounted for by that axis divided by the largest amount of variance accounted for by any axis (always axis I), i.e., axis I will always have a constant of 1.0. This procedure is essentially equivalent to weighing each axis by the fraction of total variance accounted for by all axes under consideration. The ratio between the weighing constants will be the same by either method of calculation and the relative adjustment of intersample distances will be the same. Once the coordinates have been appropriately weighed, the rejection envelope can be calculated.

There were two major approaches taken in defining the rejection envelope. The first was to calculate the three-dimensional distances from each

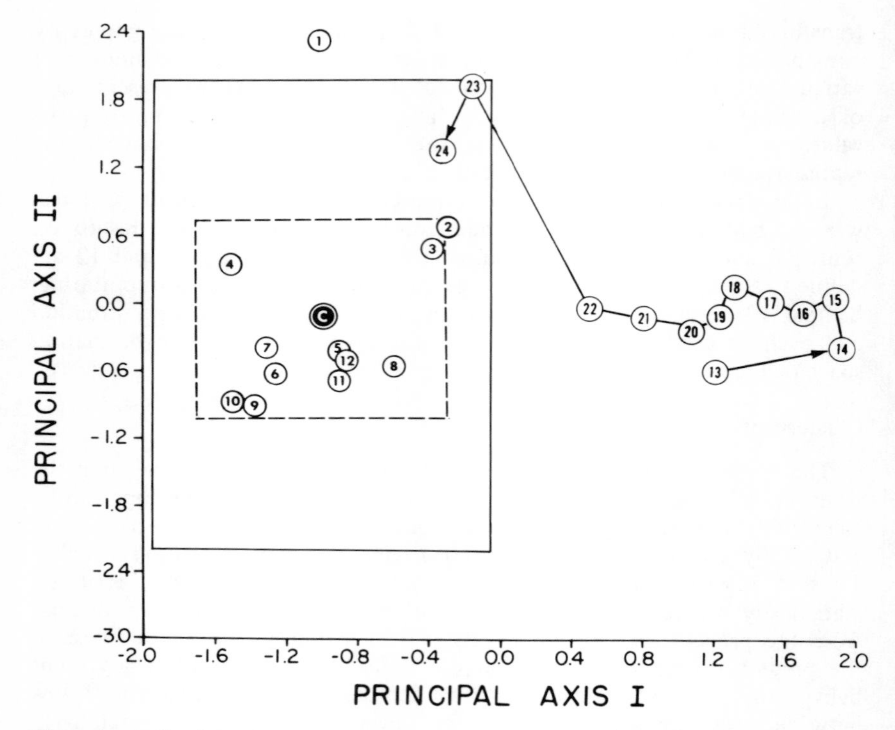

Figure 1. Ordination of a benthic infaunal community after natural defaunation. Samples 1 through 12 represent preperturbation and Samples 13 through 24 represent postperturbation (see text for details).

preperturbation sample to the centroid of the cluster. Using the parametric approach outlined below, a spherical rejection envelope could be calculated. Since this method ignores cluster shape, it was not surprising that the results were confusing. For this reason, the spherical model was rejected.

Although it is theoretically possible to standardize the axes in such a way as to make the spherical model more reasonable, a more direct approach was taken. The set of coordinates of the preperturbation samples on any axis can be characterized by their mean and standard deviation. By rearranging the "t" test for a single observation [Simpson et al. 1960], the following expression results:

$$x_c = \overline{X} \pm \frac{|t_{(df=N-1:\alpha = 0.05)}| \times S}{\sqrt{\frac{N}{N+1}}}$$

By substituting the mean $\overline{X}$, standard deviation S, the number of samples N for any axis and the appropriate t value, the critical values (above and below the mean) which will result in a rejection of the null hypothesis can be calculated. The pair of these critical values for the three axes represent the coordinates of the corners of the 95% rejection polygon. A rejection polygon more closely approximates true cluster shape than any arbitrarily selected geometric solid. The box of solid lines in Figure 1 represents the rejection polygon for the first two axes of the example presented here. If the coordinate of a postperturbation sample falls between the critical values for any axis, the postperturbation sample is within cluster for that axis. If the sample is within cluster on all axes, then the postperturbation point is statistically indistinguishable from the preperturbation state and the community can be said to have recovered.

The distribution of points within the cluster (Figure 1) along axis II appears to be bimodal and is decidedly not normal. Since the test outlined above is based on parametric statistics, the data violates the basic assumption of normality. To counter this legitimate objection, a nonparametric analog to the above procedure was created.

Along any axis, the distance between two coordinates can be represented as a line segment. The line segments between all possible pairs of preperturbation samples for a given axis can be calculated and form the preperturbation data set. In Figure 2, the preperturbation sets for axes I and II are represented by the set of bold line segments parallel and closest to those axes respectively. For any given point on the same axis, line segments between that point and the coordinates of the preperturbation samples can be calculated. If the point is the coordinate of a postperturbation sample, the set of line segments form the postperturbation data set. The postperturbation data sets are represented in Figure 2 by the sets of thinner lines parallel and further from the respective axes. By ranking the line segments of pre- and postperturbation data sets together and performing a Mann-Whitney "U" test [Siegel 1956], the null hypothesis that there is no difference between the data sets can be tested. If a point lies outside the cluster, as does the postperturbation point on axis 1 in Figure 2, the line segments of the postperturbation set will be longer than those of the preperturbation data set and the null hypothesis will be rejected. Conversely, if the point should lie within cluster, as it does along axis II, the sets of line segments will be statistically indistinguishable. The appropriate test is then a one-tailed Mann-Whitney "U" test.

To determine the rejection polygon, an iterative procedure is used. A point is arbitrarily chosen remote to the centroid coordinate (four times the standard deviation away from the centroid coordinate for a given axis). That point is then tested nonparametrically as outlined above. If it is found to be outside of cluster, a second point is chosen at half the distance from the first point to the centroid coordinate. If the first point is found to be within cluster, the second point is placed further from the centroid coordinate by a similar algorithm. By repeating the process until some arbitrary small distance (here set to 0.0001 axis units) no longer exists

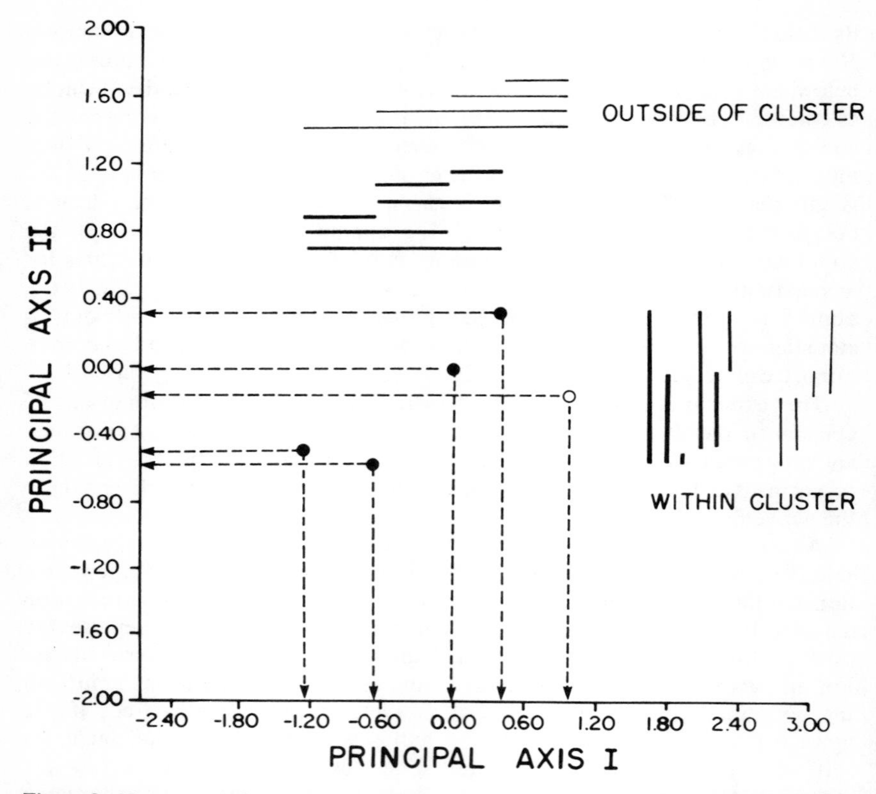

Figure 2. Nonparametric statistical method of determining whether a point is within cluster by creation and comparison of intersample line segments (see text for details).

between the first and second points, the nonparametric equivalent of the critical values of the rejection polygon for the axis is generated. The procedure is repeated for each axis and the three-dimensional rejection polygon is created. The dashed line in Figure 2 represents the nonparametric rejection polygon for axes I and II for the data set examined here.

Distance to the Rejection Envelope

A line can be generated from each postperturbation point in three dimensions to the centroid. Once the rejection polygon is defined, the coordinates of the point at which the line between the postperturbation point and the centroid pierces the nearest surface of the polygon can be determined. By application of an extended Pythagorean theorem, the distance between the postperturbation point and the pierce point can be calculated. If the postperturbation point is found to be within the cluster on all axes, a line is projected from the centroid, through the point and back to the nearest surface of the polygon. The pierce coordinates are determined and the distance between the postperturbation point and the pierce point is set as a negative to indicate that the distance is back to the surface of the polygon. In the event that the postperturbation point should fall exactly at the centroid, the distance would be the perpendicular to the nearest surface of the polygon.

The reason for the three-dimensional limitation of the technique lies in the calculation of the pierce coordinates. There are six critical zones and six critical lines defined for each axis. Therefore, for three axis there are 1728 possible locations for any point relative to the rejection polygon and the centroid. In its present form, the relative location of each postperturbation point is found by a hierarchical search procedure. Expansion to four or more axes would greatly complicate the programming and dramatically increase the computing cost of the analysis.

Once the distance from each postperturbation point to the nearest surface of the polygon ("distance to recovery") is known, these data can be plotted against time. In Figure 3, the solid line represents the parametric result and the dotted line represents the nonparametric result. As was evident from the relative volumes of the polygons in Figure 2, the nonparametric analysis is less likely to show recovery. However, there is no real choice between the two types of analyses. In that the parametric assumptions are not likely to be met, the nonparametric approach is the valid statistical procedure.

Availability of Programs

Principal Coordinate Analysis is supported by many computer centers and is available as part of a large package [Bloom et al. 1977]. The program for the calculation of the rejection polygon, the pierce coordinates, and the plotting data was written in FORTRAN IV for an Amdahl 470 (compatible with IBM 360/370). A listing of the program is available from the author upon request.

DISCUSSION

As can be seen in Figure 3, the community analyzed here does show a distinct trend towards recovery (as it should given the definition of the preperturbation cluster as being really the second year of recovery). There is an initial divergence from the preperturbation state (most likely due to the presence of several normally rare species which were unusually numerous in the second month after perturbation). There is a gradual fall towards recovery up to month 11 and then a minor divergence at month 12. The analysis thus appears to be a meaningful way to express recovery. The speed of recovery is expressed by the general slope of the falling line. If there

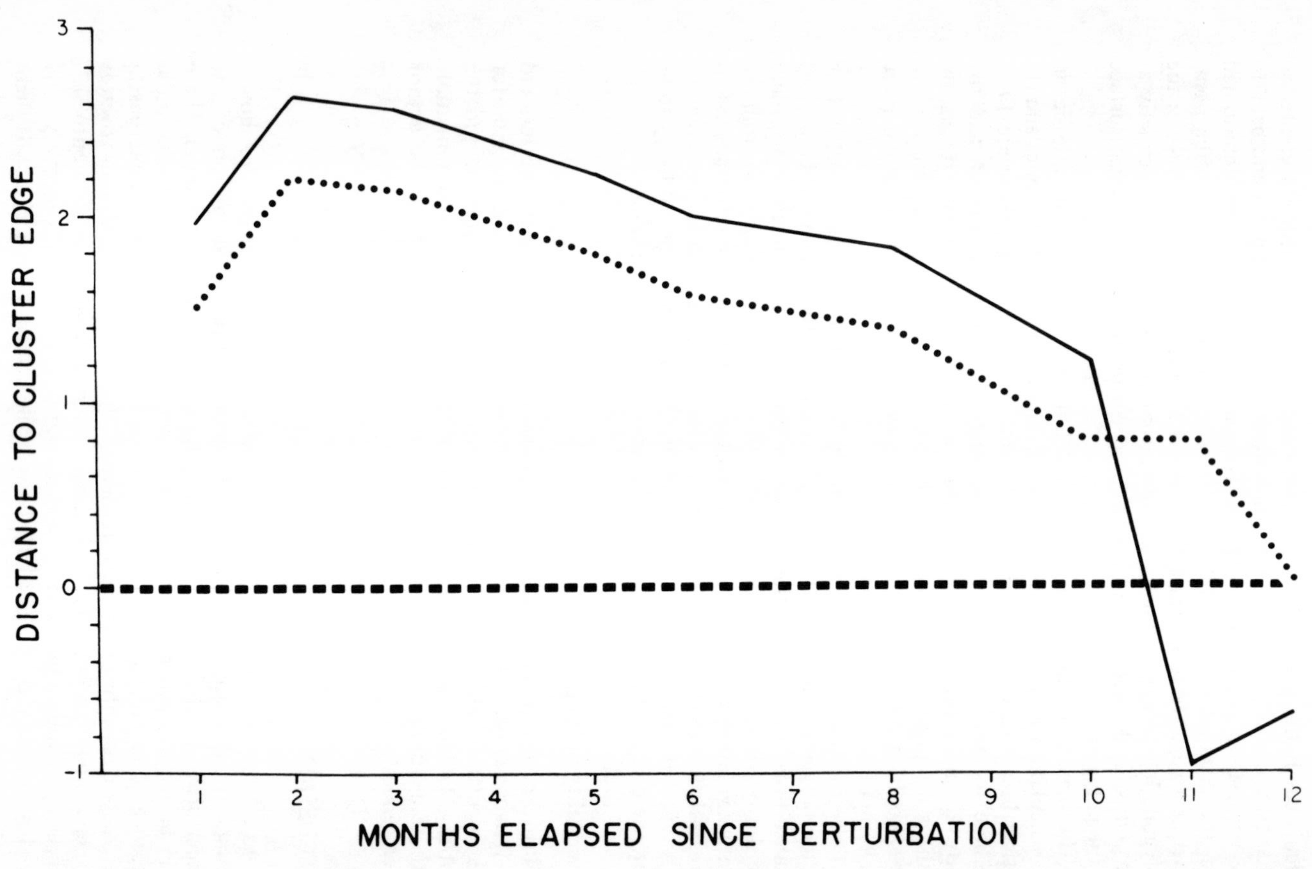

Figure 3. Recovery analysis of a benthic infaunal community after natural defaunation using a square-root transformation with parametric and nonparametric statistical techniques (see text for details).

were multistable points, the line would plateau away from the preperturbation cluster, i.e., a second cluster would form in ordination space. If a series of successional stages were encompassed within the general recovery path, a staircase effect would result. Thus the slope and shape of the line can express meaningful concepts.

The analysis can also be profitably used with experimental data. Control data can form the "preperturbation" cluster while the experimental data can be regarded as the "postperturbation" data. The response of an experimental community to perturbation could thus be monitored and the impact of the experimental perturbation with regard to community alteration can be quantified.

The limitations of the proposed technique should be kept firmly in mind. Comparable before-and-after quantitative data are absolutely necessary. Although the first three axes of a PCOR should account for the majority of variation in the data, for many data sets they do not. The extension of the analysis to more than three dimensions is therefore desirable. On theoretical grounds, a principal component analysis (PCA) would seem to be superior in that the axes' rotation could be solely based on the preperturbation data. The postperturbation samples could then be introduced into the axes system after a coordinate transformation [Sneath and Sokal 1973]. In this manner, the rejection polygon would be only influenced by the preperturbation data and would overestimate the actual cluster shape by the smallest amount possible. In a trial run, however, a PCA obliterated any meaningful pattern in the data. Presumably, the noted space-distortion of a PCA [Rohlf 1968] creates more problems than are cured by excluding the postperturbation data from consideration in the formation of the rejection polygon.

In that data transformation is not specified in running a PCOR, a valid concern is how much variation in the final result can be generated by using different transformations. In Figure 4, the results of analyzing the same data set using no transformation, a frequency standardization (the number of individuals of a species at a station divided by the total number of individuals of all species at that station), logarithmic, square-root and cube-root transformations [Boesch 1977] are presented. The various data transformations do little to alter the general pattern of recovery and the last three transformations yield virtually identical results. The analysis thus appears to be quite robust and is relatively insensitive to even major alterations in the initial data matrix.

The technique presented here should be of value in assessing environmental changes, in monitoring of communities under chronic and acute perturbations and in helping analyze experimental perturbation systems. Alterations can be made in the various analytical subsections of the technique such as ordination, the number of axes used, the nonparametric test utilized and the method of calculating distance to recovery. However, the basic framework of representing intersample dissimilarities as projected distances, defining a rejection region based on the preperturbation data, and expressing the distance to recovery through time is not only intuitively simple but places questions of community recovery in a testable statistical frame.

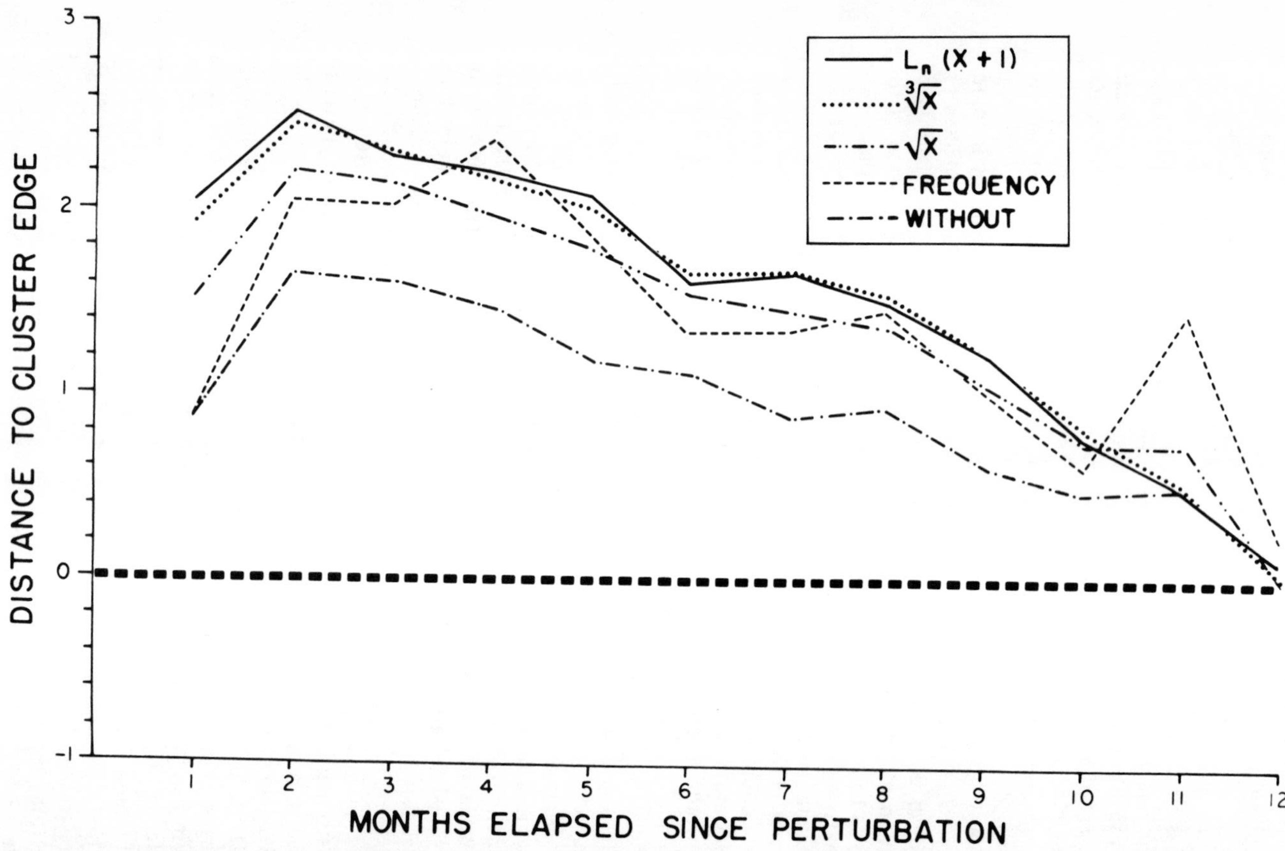

Figure 4. A comparison of recovery analyses of a benthic infaunal community after natural defaunation using a nonparametric statistical technique and five different data transformations.

CONCLUSIONS

A multivariate, nonparametric statistical and geometrical technique that converts before and after perturbation data into a meaningful representation of the distance to recovery through time is described. An example is analyzed and the limitations and areas for improvement are specified. The technique can be used in both descriptive and experimental systems and can be used to quantify recovery, successional transitions and the existence of multistable states.

ACKNOWLEDGMENTS

I thank Dr. Joseph L. Simon for making the data set available, Dr. Stuart Santos for providing a sounding-board, Mr. Robert Smith for his cogent criticisms and suggestions and the University of Florida Department of Zoology for making computer and support funds available.

REFERENCES

Allen, T. F. H., S. M. Bartell and J. F. Koonce. "Multiple stable configurations in ordinations of phytoplankton community change rates," *Ecology* 58:1076–1084 (1977).

Boesch, D. F. "Application of numerical classification in ecological investigations of water pollution," Spec. Sci. Report 77, VIMS; Ecol. Res. Series EPA-600/3-77-033 (1977), 115 pp.

Bloom, S. A., S. Santos and J. G. Field. "A package of computer programs for benthic community analyses," *Bull. Mar. Sci.* 27:577–580 (1977).

Bloom, S. A., J. L. Simon and V. D. Hunter. "Animal-sediment relations and community analysis of a Florida estuary," *Mar. Biol.* 13:43–56 (1972).

Dauer, D. M., and J. L. Simon. "Repopulation of the polychaete fauna of an intertidal habitat following natural defaunation: species equilibrium," *Oecologia* 22:99–117 (1976).

Holling, C. S. "Resilience and stability of ecological systems," *Ann. Rev. Ecol. Syst.* 4:1–25 (1973).

MacArthur, R. H., and E. O. Wilson. *The Theory of Island Biogeography* (Princeton, NJ: Princeton University Press, 1967), 203 pp.

Rohlf, F. J. "Stereograms in numerical taxonomy," *Syst. Zool.* 17:246–255 (1968).

Siegel, S. *Nonparametric Statistics for the Behavioral Sciences* (New York: McGraw-Hill Book Company, 1956), 312 pp.

Simberloff, D. S. "Equilibrium theory of island biogeography and ecology," *Ann. Rev. Ecol. Syst.* 5:161–182 (1974).

Simpson, G. G., A. Roe and R. C. Lewontin. *Quantitative Zoology* (New York: Harcourt, Brace & World, 1960), 440 pp.

Sneath, P. H. A., and R. R. Sokal. *Numerical Taxonomy* (San Francisco: W. H. Freeman & Company, 1973), 573 pp.

Sutherland, J. P. "Multiple stable points in natural communities," *Am. Nat.* 108:859–873 (1974).

Whittaker, R. H. *Ordination and Classification of Communities* (The Hague: W. Junk B. V. Publishers, 1973), 737 pp.

CHAPTER 7

THE 'ŌHI'A DIEBACK PHENOMENON IN THE HAWAIIAN RAIN FOREST

Dieter Mueller-Dombois
Department of Botany
University of Hawaii at Manoa
Honolulu, Hawaii 96822

INTRODUCTION

A widespread and significant tree dieback was discovered a decade ago [Mueller-Dombois and Krajina 1968] in the natural rain forest on the island of Hawaii. Subsequently, an aerial photo analysis was made of the dieback phenomenon using three successively prepared photographic sets made in 1954, 1965 and 1972. It was concluded [Petteys et al. 1975] that the native rain forest was rapidly declining. The decline was described as a "severe epidemic," and a prediction was made that the native rain forest would be eliminated in 15–25 years if the present rate of damage continued. This prediction implied the assumption that the native rain forest was struck by a newly introduced disease.

Intensive disease research was begun in 1972 by the U.S. Forest Service. Several potential pathogens were isolated, among them *Phytophthora cinnamomi*, a soil and root fungus which causes large-scale native forest decline in West and Southeast Australia [Weste and Taylor 1971; Podger 1972; Weste and Law 1973].

Consecutively with the disease research, the author began to study the dieback phenomenon with three graduate students (James D. Jacobi, Ranjit G. Cooray, N. Balakrishnan) in 1974 under the general hypothesis that the dieback may be a recurring natural phenomenon in primary succession. Three observations led me to assume that the dieback was a natural phenomenon.

First, during my initial observations in 1965 I noticed that the dieback occurred mostly on inundated or poorly drained sites. Second, searching with my students through the literature, we came upon the "Maui Forest Disease," an 'ōhi'a rain forest dieback that had occurred in the early part of this century on the island of Maui [Lyon 1909]. Despite several years of intensive research, no pathogen could be detected for the Maui forest disease. No biotic agent was found to be involved [Lyon 1919]. Third, despite claims by some foresters [Petteys et al. 1975] that 'ōhi'a trees and other tree species were dying in all age classes, I could never find any appreciable death or dying of the undergrowth. Only upper-story 'ōhi'a trees were dead or dying in large quantities (Figure 1).

A short description of the area and a brief explanation of the author's field approach and current results; a summary of the major findings to date; and a dieback theory and new working hypothesis (based on Mueller-Dombois et al. 1977) will be given.

AREA

The air photo analysis relates to an 80,000-ha territory (i.e., an area of about 16 × 50 km) on the east slopes of Mauna Kea and Mauna Loa, between 610 and 1830 m (2000–6000 ft) altitude (Figure 2). Throughout most of this territory, the natural forest is dominated by only one tall tree species, the native *Metrosideros collina polymorpha*, locally called 'ōhi'a-lehua. A second tall tree species, the native *Acacia koa*, occurs in a narrow

Figure 1. The 'ōhi'a dieback along a study transect taken in fall 1965 (photo by D. Mueller-Dombois).

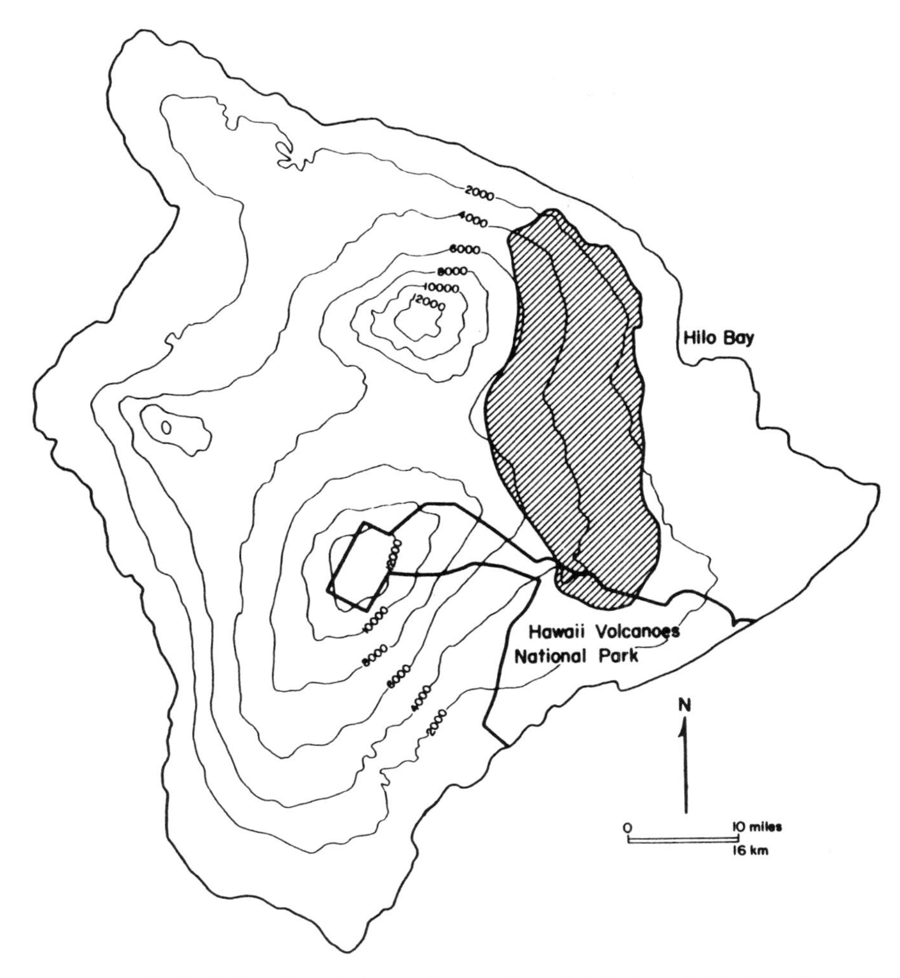

Figure 2. Island of Hawaii and the study area (cross-hatched area). Contour lines at 2000-ft intervals.

belt in the upper altitudinal fringe area of this territory and also is more localized in the lower altitudinal fringe area on the east slope of Mauna Kea. In addition to the monodominant 'ōhi'a, about 20 other woody plant species occur, usually lower-growing trees and arborescent shrubs. These are associated with the 'ōhi'a in various proportions. A second dominant and widely distributed, 'ōhi'a-associated, plant life form is the tree ferns, particularly *Cibotium glaucum* (hāpu'u). Typically, the tree ferns form a distinct second layer at about 2–3 m, with the 'ōhi'a trees reaching above to 10–20 m.

Depending on the density and activity patterns of feral pigs the forest floor may be covered by a dense growth of native, herbaceous ferns, or it may be very sparsely covered by plants.

Other introduced mammals include the roof rat (*Rattus rattus*) and the mongoose (*Herpestes auropunctatus*) and very rarely a feral cat or dog. The native fauna consists of a great number of species of arthropods and tree snails and of several species of forest birds, most of them belonging to the honeycreeper family (Drepanididae).

The climate of the area is continuously humid, with monthly rainfalls almost always in excess of 100 mm. The annual rainfall varies from about 1.8 m in the south part of the study area on Mauna Loa to 7.5 m in the middle-area on Mauna Kea. The mean annual temperature at 610 m (2000 ft) is about 20°C, at 1830 m (6000 ft) it is about 15°C. Temperature is equable throughout the year, i.e., summer and winter temperatures differ by only a few degrees (about 4–6°C). Daily temperature ranges are often somewhat greater (8–10°C). Therefore, the forest fits into the concept of tropical montane rain forests on a worldwide scale [Ellenberg and Mueller-Dombois 1967].

The substrate is volcanic. On Mauna Kea it consists mostly of deeply weathered ash of heavy clay texture. Between the two mountains and on Mauna Loa, the substrate consists mostly of basaltic lava flows (pāhoehoe and 'a'ā lava). Depending largely on geologic age, location and drainage, the lava flows are covered with a shallow layer (10–50 cm) of woody peat (histosols) or a mixture of ash and black muck.

METHODS

We began with a detailed belt-transect analysis in a severe dieback area. The transect was 6 m wide and 1.3 km long. On the transect we enumerated all ohia trees in 3 × 5 m subplots by their sizes from small seedlings to mature trees and snags. In addition, we prepared species checklists and quantitative evaluations by the Braun-Blanquet method [Mueller-Dombois and Ellenberg 1974] for every subplot along the transect. Soil pits were dug at every obvious change in substrate, profiles were described and soil samples taken for further analyses. Another long transect was done similarly in a "healthy" forest.

Following the transect analyses we prepared a large-scale vegetation map from the 1972 color-air-photos, with intensive ground checking. Ground checking involved the establishment of relevés (vegetation samples) in many of the important air-photo patterns (with and without dieback) and in different geographic locations of the study area. So far, we established 40 relevés, each 400 m^2 in size.

In each relevé, a total floristic analysis was made using the Braun-Blanquet method. One or two soil pits were dug, described in detail and samples taken. The soil surface of each relevé was analyzed for its drainage condition by the point-sampling method along predetermined transects and the topographic setting and the microtopography were described for each sample site. Full enumeration analyses of all 'ōhi'a population members from seedling to snag were made in each relevé, as was done along the transects.

In addition, further studies were done by Dr. W. Ko and S. C. Hwang on the biology of *Phytophthora cinnamomi* and its possible relationship to the dieback.

RESULTS

The soil profile and drainage analyses done in each relevé allowed us to prepare a preliminary habitat classification. Two habitat series were recognized, one with shallow soils (<50 cm fine soil or organic deposit over lava rock substrate), another with deep soils (>50 cm). In addition, five drainage classes were recognized. These ranged from well-drained to poorly drained in the shallow soil series and from well-drained to permanently water-soaked in the deep soil series.

Surprisingly, 'ōhi'a dieback stands were found in all moisture regimes and soil-depth ranges. Earlier, I had overlooked the dieback on well-drained sites. The occurrence of the dieback across so many soil variations gave some support to the disease hypothesis. However, the dieback patterns differed in relation to these site variation. Dieback over large areas, continuous over several hectares, was found only on poorly drained sites. On deep, boggy soils there was a salt-and-pepper pattern of dieback, i.e., healthy crowned trees and snags occurred intermixed. On shallow, poorly drained soils a more continuous tree-to-tree dieback pattern occurred. This was the more typical large-area dieback, which originally drew attention to the problem. On the better drained sites, dieback was much less area-extensive. Two patterns were also found, a tree-to-tree dieback and a less complete dieback which had a salt-and-pepper appearance on air photos. The tree-to-tree dieback occurs on well-drained soils in pockets of up to one hectare, but it usually occupies still smaller areas. This pocket dieback was formerly recognized as "hotspots" or "disease centers," but wherever investigated the results were negative for disease-causing organisms. The second dieback on the better drained sites, which also resembles a salt-and-pepper pattern was found locally only, but over a larger area of about 25 km^2, on moderately well-drained, deep ash soils, which are rich in nutrients.

Our vegetation map analysis has shown that general vegetation structure patterns such as closed versus open canopy forests are not well correlated with the physical habitat types. This means that both closed and open forests can be found on nearly all moisture regimes and soil depth types. This in turn implies a dynamic nature to forest denseness instead of a site-indicator value.

Our population counts of 'ōhi'a trees and seedlings in form of density/size class have brought out an almost direct relationship between openness of the canopy and 'ōhi'a tree reproduction. In all relevés sampled so far in the tree-to-tree diebacks, on both well-drained and poorly drained sites, we found adequate 'ōhi'a reproduction. We defined "adequate reproduction" as at least 3500 seedlings and saplings from 0.1–5 m height per hectare. The only form of 'ōhi'a dieback, under which 'ōhi'a reproduction was not adequate (according to the above standard) is the salt-and-pepper dieback on the fertile ash soil. Here, the growth of tree ferns (*Cibotium* spp.) is so vigorous that the ground is almost totally shaded by them, and tree fern density is very high. The dieback seems to be resulting in a displacement of 'ōhi'a and a development of almost pure tree fern forests. Our data is insufficient on the salt-and-pepper dieback of the boggy deep-soil sites, but our preliminary observations indicate a slow decline of 'ōhi'a and a displacement of trees by shrubby forms of 'ōhi'a.

SIGNIFICANT FINDINGS

Significant findings contributed to the 'ōhi'a dieback problem by our study can be summarized as follows:

1. The dieback is manifested only on canopy trees.
2. There is a site relationship with regard to the dieback pattern, i.e., small-area "tree-to-tree" dieback occurs in stands on well-drained sites, while large-area "tree-to-tree" dieback occurs on poorly drained sites. Both occur on shallow soils. There are also "salt-and-pepper" dieback patterns on deep-soil habitats, particularly on moderately drained, fertile ash soils and on boggy soils. Thus, we can distinguish at least four site-related dieback patterns.
3. Regardless of soil moisture regime, all 18 tree-to-tree dieback sample stands examined showed significant 'ōhi'a reproduction ($>$3500 seedlings or saplings/ha). The four salt-and-pepper dieback sample stands examined showed inadequate 'ōhi'a reproduction. All nine closed-canopy or "healthy" stands examined showed only few small seedlings and lacked taller ones. This indicates clearly that the tree-to-tree dieback is resulting in rejuvenation of the forest and not, as originally suggested, in a forest decline.

According to the air photo analysis done by the U.S. Forest Service in 1975 [Petteys et al. 1975], 120 ha were recognized as showing severe decline on the 1954 air photo set, 16,000 ha on the 1965 set and 34,500 on the 1972 set. On this basis the conclusion was drawn that the remaining 'ōhi'a forest in the 80,000-ha study area would succumb to the epidemic in another 15 to 25 years, if existing conditions continued.

Clearly, the conditions did not continue, because by extrapolation there should only have been approximately 5000 ha left in 1978. Instead the area with "healthy" forest is still extensive. The author is working on a more detailed mapping analysis of the dieback patterns, but by estimation severe decline still has not spread over more than 50% of the area.

The initial air photo interpretation was undoubtedly exaggerated. The air photo analysis may have suffered because of failure to distinguish between the different dieback patterns. However, there is no doubt that the tree-to-tree dieback is a rapid and spatially significant phenomenon.

As to the causal mechanism of the dieback, three findings are of significance:

1. The soil and root study of *Phytophthora cinnamomi*, done within the framework of the author's project [Hwang 1977] found that the fungus is more common at lower elevations and almost absent at higher elevations. Furthermore, it is a good saprophyte and shows no casual relationship to the 'ōhi'a dieback. In other words, healthy stands at lower elevations (around 610 m) showed the fungus in the soil and some roots, dieback stands at higher elevations (around 1220 m) lacked the fungus in soil and roots.

2. Insect and disease research done by the U.S. Forest Service has revealed that among several disease-causing candidates, such as *Armellaria mellea*, *Pythium vexans* and others, *Phytophthora cinnamomi* was considered the most likely cause of the 'ōhi'a decline. The clearest finding (Dr. Ko, personal communication) is that *P. cinnamomi* has a relationship to soil moisture. It occurs only in moist, but not in either well-drained or very poorly drained sites. Dieback, however, occurs on all of these sites. It is currently suspected that the fungus may have a secondary or contributory role to dieback on moist sites.

Among insects, only the native host-specific wood borer *Plagithmysus bilineatus* was suspected as a causative agent. Intensive research over the last several years has shown that it is not causative, but that it may have a "hastening effect" on the dieback, once the trees are physiologically weakened from another cause [Papp et al. 1979].

3. A few plots with dieback trees were treated with NPK fertilizer. The declining tree crowns recovered with foliage [Kliejunas and Ko 1974]. This fertilizing experiment indicated that the 'ōhi'a dieback was related to nutrient deficiency in the soil.

When originally discovered, this fertilizer response seemed puzzling. Nutrient starvation of forest stands in natural environments does not seem a commonly reported phenomenon.

CONCLUSIONS

The six important findings about the rain forest dieback problem in Hawaii can be fitted into a new dieback theory.

It is almost certain that the cause is not a newly introduced insect pest or disease-causing organism. This was initially suspected not only because of the large spatial extent of the dieback, but also because we are geared to think that island ecosystems are fragile. Even indigenous biotic agents are not involved as initiating causes of the dieback. The dieback then is clearly a natural phenomenon in the dynamics of this rain forest, which is initiated by an abiotic environmental cause in combination with the ecophysiological properties of 'ōhi'a.

Our current working hypothesis is that the dieback is initiated by a climatic instability which becomes effective through the soil moisture regime under certain conditions of forest stand maturity. As climatic instability we suspect excessive rainfall in certain years, which may flood and drown out the root systems of the taller trees on poorly drained sites. Dieback under these conditions may be associated with iron toxicity. On well-drained sites we expect the opposite climatic instability to initiate the dieback syndrome. A certain "low" in the rainfall pattern may manifest itself as soil drought. Trees, which are already surviving under nutrient starvation, particularly when this is aggravated by intra- or interspecific competition, may suddenly succumb and die. Lightning discharge can be another cause.

We have found that the tree-to-tree diebacks on both well-drained and poorly drained soils are associated with rejuvenation of the same tree species. The dieback thus has become a successful mechanism to maintain an essentially shade-intolerant pioneer species as the structure-forming dominant in the course of primary succession.

REFERENCES

Ellenberg, H., and D. Mueller-Dombois. "Tentative physiognomic-ecological classification of plant formations of the earth," Rübel, Zürich, *Ber. geobot. Inst. ETH Stiftg*. 37:21–55 (1967).

Hwang, S. H. "'Ōhi'a decline: the role of *Phytophthora cinnamomi*," CPSU (University Hawaii, Botany Department) Tech. Rep. No. 12 (1977), 71 pp.

Kliejunas, J. T., and W. H. Ko. "Deficiency of inorganic nutrients as contributing factor to 'ōhi'a decline," *Phytopathology* 64:891–896 (1974).

Lyon, H. L. "The forest disease on Maui," *Hawaiian Planter's Record* 1:151–159 (1909).

Lyon, H. L. "Some observations on the forest problems of Hawaii," *Hawaiian Planter's Record* 21:289–300 (1919).

Mueller-Dombois, D., J. D. Jacobi, R. G. Cooray and N. Balakrishnan. "'Ōhi'a rain forest study, final report," CPSU (University Hawaii, Botany Department) Tech. Rep. No. 20 (1977), 117 pp. + 3 map sheets.

Mueller-Dombois, D., and H. Ellenberg. *Aims and Methods of Vegetation Ecology* (New York: John Wiley & Sons, Inc., 1974), 547 pp.

Mueller-Dombois, D., and V. J. Krajina. "Comparison of east-flank vegetations on Mauna Loa and Mauna Kea, Hawaii," *Recent Adv. Trop. Ecol.* 2:508–520 (1968).

Papp, R. P., J. T. Kliejunas, R. S. Smith, Jr., and R. F. Scharpf. "Association of *Plagithmysus bilineatus* (Coleoptera: Cerambycidae) and *Phytophthora cinnamomi* with the decline of 'ōhi'a-lehua forests on the island of Hawaii," *Forest Service* 25(1):187–196 (1979).

Petteys, E. Q. P., R. E. Burgan and R. E. Nelson. "'Ōhi'a forest decline: its spread and severity in Hawaii," USDA Forest Service Res. Paper PSW-105, Berkeley, CA (1975), 11 pp.

Podger, F. D. "*Phytophthora cinnamomi*, a cause of lethal disease in indigenous plant communities in Western Australia," *Phytopathology* 62:972–981 (1972).

Weste, G., and C. Law. "The invasion of native forests by *Phytophthora cinnamomi*. III. Threat to the National Park, Wilson's Promontory, Victoria," *Aust. J. Bot.* 21:31–51 (1973).

Weste, G., and P. Taylor. "The invasion of native forests by *Phytophthora cinnamomi*. I. Brisbane Ranges, Victoria," *Aust. J. Bot.* 19:281–294 (1971).

CONCLUSION

To present a synopsis of the recovery process for damaged ecosystems based on a one-day symposium, however excellent the papers, would be presumptuous. However, some important messages to policy-makers and management personnel are evident.

Ecosystems damaged by societal activities may reacquire lost ecological qualities by natural processes.

Ecosystems dependent upon periodic natural disturbances (e.g., floods and fires) may be markedly changed if these disturbances are controlled by man.

Damaged ecosystems may: (1) return to their original condition; (2) be rehabilitated to a condition which includes some of the original characteristics and some new characteristics considered to be beneficial to man; (3) be "enhanced" by management techniques to an improved but different condition from the original; or (4) remain in a damaged state with consequent loss of amenities.

A much better understanding of the various processes (e.g., succession, recovery, rehabilitation and enhancement) is essential.

Present understanding of succession is inadequate to predict the series of biological events that will occur based on present and past conditions.

Disequilibrium may be followed by a return to the original condition or by establishment of a new equilibrium condition. A basis exists for estimating which is most probable.

Management practices are available to enhance the recovery of a displaced plant species.

Damaged ecosystems of considerable size (e.g., the Great Lakes) may be rehabilitated to a condition ecologically superior to the present despite formidable obstacles. Such a possibility should not be interpreted as a license to destroy healthy systems but rather a call for rehabilitation of those already damaged.

Quantification of recovery is difficult, but not impossible. This procedure may be extremely important if those legally responsible for damaging ecosystems are required to restore or rehabilitate them.

An inappropriate intervention may occur if natural cycles, including catastrophic events (e.g., 'ōhi'a rainforest), are not understood.

If comparable pre- and postperturbation data sets are obtained, statistical packages are available which assess the extent to which an ecosystem has recovered.

Man-induced disturbances should be examined a priori so that the implications of such actions may be thoroughly understood and damage to the system may be mitigated.

Future policies and actions should not be limited by past dogmas and current schools of thought but should be based on research on the impact of natural and manmade perturbations.

The ultimate success in rehabilitating ecosystems depends on the cooperation of the political, legal and ecological agencies involved.

Natural processes work slowly, but well-planned management programs can quicken the recovery of a damaged system.

This small volume relates other conclusions and messages not recounted here, and a one-day symposium necessitated limitations. If any chapter calls attention to a problem that has not heretofore received the attention it deserves, the book will have served its purpose. Whether it does so by inspiration or irritation is irrelevant.

INDEX